New Herb Bible

GROWING • COOKING • HEALTH REMEDIES
COSMETIC USES

4-26-14
Happiness always
Love,
Mom
OX

New Herb Bible

GROWING • COOKING • HEALTH REMEDIES
COSMETIC USES

Caroline Foley • Jill Nice • Marcus A. Webb •

METRO BOOKS
New York

METRO BOOKS
New York

An Imprint of Sterling Publishing
387 Park Avenue South
New York, NY 10016

METRO BOOKS and the distinctive Metro Books
logo are trademarks of Sterling Publishing Co., Inc.

Copyright © 2001 Quintet Publishing Limited

This 2013 edition published by Metro Books by
arrangement with Quantum Books.

Managing Editor: Diana Steedman
Art Director: Sharanjit Dhol
Designer: Isobel Gillan
Creative Director: Richard Dewing
Publisher: Oliver Salzmann

ISBN 978-1-4351-4783-6

For information about custom editions, special sales,
and premium and corporate purchases, please contact
Sterling Special Sales at 800-805-5489 or specialsales@
sterlingpublishing.com.

Manufactured in Singapore by
Universal Graphics Pte Ltd
Printed in China

2 4 6 8 10 9 7 5 3 1

www.sterlingpublishing.com

This book contains recipes using herbs and other
ingredients that may cause an allergic reaction in some
individuals. Care should be taken in handling all
ingredients. Neither the authors nor the publisher is
responsible for any adverse effects or consequences
allegedly resulting from the use of the recipes, formulas,
or other information or suggestion in this book.

contents

introduction *8*

herb gardens *10*

herb directory *46*

herbs for cosmetics *104*

medicinal herbs *142*

cooking with herbs *166*

resources and credits *218*

index *219*

introduction

Most modern research on herbal remedies simply confirms what has been observed for centuries, that herbs play an important role in medicinal, aromatherapeutic, cosmetic, and culinary practice. People have made use of herbs since the beginning of civilization, when they depended on them for food, medicine, and preservatives. As a result, herbs attained a high importance in religious rites and festivals and many superstitions grew up around them.

Ancient herbals have absorbed readers since Dioscorides wrote in the first century AD his *De Materia Medica*, describing plants for their healing qualities. The Greek physician observed their growth and how to gather and store them. The earliest existing copy of his great work, written and illustrated in the sixth century, the *Codex Vindobenensis*, is housed in Vienna.

Other ancient manuals identifying medicinal plants came to include the culinary as well as their medicinal qualities, and many were published in medieval Europe. During the fifteenth century von Megengerg's *Buch der natur, Book of Nature* was printed with the first known woodcuts for botanical illustrations, and by the sixteenth century, as explorers brought plants back, the emphasis on accuracy became more and more important. By the time of John Gerard's *Herball* of 1597, herbals were more than simply catalogs of plants, for his famous work incorporated practical detail and knowledge acquired from growing herbs, newly arrived from the Mediterranean and the New World, in his London garden.

Alongside the publication of genuine herbals were other works concerned with the medical theory and use of plants to cure human ailments. Nicholas Culpeper's *A Physicall Directory* set out the medicinal plants and defined their uses. He was a follower of a form of natural healing known as the Doctrine of Signatures, which teaches that like cures like: for example, that red flowers are most likely to cure disorders of the blood.

Knowledge of the use of herbs spread throughout Europe and was taken to North America by the early settlers. It was the Shaker communities that first grew and sold herbs as a successful commercial enterprise. The Native Americans contributed greatly to health and survival using only natural agents; the French physician Nostradamus used natural ingredients to help people suffering from the plague; traditional Chinese medicine had its approach firmly rooted in the therapeutic properties of plants.

In the west, with the advent of chemical medicine and modern storage and preserving techniques, the use of herbs declined and people lost their awareness of their natural benefits. Yet the pendulum has swung back, and now there is unprecedented revival of interest in these versatile plants. Natural practitioners believe that treatment with chemical drugs has serious disadvantages, and our bodies are not designed to absorb inorganic matter. People are again getting in touch with the advantages of natural produce and being ever more adventurous with the use of a large variety of herbs.

Herb gardens

HERBS ARE STEEPED IN HISTORY AND ROMANCE. EVEN THEIR NAMES read like poetry – the apothecary's rose, rosemary, honeysuckle, lady's mantle, and marigold. They are a delight to grow, not only for their diverse flowers, scents, and forms, but because they are the simplest group of plants to cultivate. Their thousands of years' adaptation to tough conditions in the wild has made them born survivors. With their robust constitutions, herbs rarely suffer ailments. In fact many act as guardians protecting the plants around them. The shrubby Mediterranean herbs, rosemary and lavender, make a cloud of powerful essential oils around themselves confusing pests that go by smell. The pungent reek of rue and bitter wormwood drives away slugs and moles. The daisy flowers on many draw in the useful predators, ladybugs, and lacewings that keep aphids at bay.

Where there are herbs there will be bees and butterflies. Science, which has changed many flower forms out of recognition in an attempt to make ever more glamorous new plants, has had less success with herbs. Most remain close to the wild with both the "simple" flowers and the scent which draws in wildlife like a magnet.

With a few notable exceptions, herbs are not flamboyant but set off the lead players in the herbaceous border. There are herbs for every situation. Most of the favorite garden-worthy herbs originate from the Mediterranean coast where summers are hot and dry and winters mild. Given sunshine, protection from cold winds, free-draining soil, and temperatures that do not drop much below freezing, they are easy to grow and have uses not found in other plants. ■

The culinary garden

Most culinary herbs live happily in containers with the advantage that they can be shuffled around according to their needs and to create the best visual effect.

There are great advantages to growing them in this way. You can juggle them, bringing those in flower to the front. Containers allow you to give any particular plant the right soil and drainage. You can control the amount of sunshine or shade, water and feed and there is no worry about soil-borne pests. Mints, tarragon, and other herbs which colonize and spread in an open space are kept under control.

Almost anything can be turned into a container as long as you ensure that it has drainage holes. Old galvanized buckets, sinks, water butts and tanks can be revamped. A wheelbarrow filled with pots has a rustic air. Fruit and vegetable crates, baskets and trugs have charm. Line them with heavy-duty polythene and paint with wood preservative to make them last longer

A Versailles tub or half-barrel makes a home for shrubs and trees. You cannot go wrong with clay pots. If they are new, soak them for a few hours to prevent them from drying the roots. A parsley pot is attractive but watering is difficult due to the holes in the sides. The way round this is to sink a piece of open ended pipe vertically into the compost.

If weight is a consideration on a balcony or roof, use fiberglass or plastic containers. Paint them a dark color so that they recede, or disguise them by wrapping them around with rushes tied with wire.

Right Herbs for cooking and scent are conveniently placed by the kitchen door. Lavender, parsley, bay, rosemary, and oregano make a good collection.

General care

Plants in containers have limited soil and are more dependent on you to keep them watered and fed than they would be in the garden. While few herbs will lose their potency if given too much fertilizer, as a general rule give them a liquid feed every two weeks through summer. Group plants with the same needs together. Even in the smallest space there are corners that are sun traps while others get the wind and shade. When gathering always leave plenty of leaves on the plant so that it can recover.

Hanging baskets

To make the best use of a small space and bring scent to nose level and delicate flowers to eye level, make use of hanging baskets. Keep in mind the weight when full of wet soil and make sure that they are securely attached.

Line the basket with sphagnum or carpet moss which can be bought from a florist or nursery. Follow this with a plastic liner, into which you punch holes to allow for drainage. If well watered it will stay mossy green all summer. If you want plants to grow through the bottom of the basket, wrap them so they don't break and thread them through the plastic layer carefully from the inside. Then fill with the lightweight compost. Hanging baskets need frequent watering in hot weather, sometimes twice a day.

Backyard containers

If you have a sunny spot on a balcony or by the back door, start with some hardy herbs to give year round structure. Given sun, drainage and shelter they need little attention.

Common sage, *Salvia officinalis*, which grows to about 28 inches is evergreen with blue flowers in summer. With its graceful drooping habit and velvety leaves, it looks elegant in a pot. There are purple-leaved varieties (the Purpurascens Group) and Icterina is a golden version. Rosemary, *Rosmarinus officinalis*, is a lovely year-round herb, flowering from spring to early summer. There are varieties with flowers every shade of blue, lilac, and pink and white. For height and narrowness try Miss Jessop's Upright, which can reach 6 feet 6 inches and has pale blue flowers. The creeping types (the Prostratus Group) will trail over a pot attractively.

A little more substantial is the bay tree, *Laurus nobilis*, which though a potential giant, takes clipping and can live for a great many years in a large container, making an architectural feature if shaped into a pyramid or ball. Bays appreciate frequent feeding and watering as the roots are near the surface.

The common or French thyme, *Thymus vulgaris*, makes a neat evergreen bush 12 inches high with lilac flowers in summer. It likes fast drainage and full sun. Do not over indulge it with fertilizer or water or it will lose flavor. Trim after

A terracotta window box, as useful as it is pretty, is planted with basil, sage, fennel, thyme, and pelargoniums.

flowering to encourage more leaf growth. A pretty kitchen variety is Silver Posie.

For a contrast in form, try the tall and airy fennel, *Foeniculum vulgare*. A hardy perennial, it can reach 5 feet and has flat umbels of yellow flowers. It will need a deep pot to anchor it and possibly some staking. Fennel likes fertile, well drained loam. Only the young growth is good for harvesting so cut off the older, outer shoots to the base to encourage new young growth. Fennel will cross-pollinate with dill and cilantro, so keep them apart.

Dill, *Anethum graveolens*, looks rather like fennel although it is not so tall at 24 to 36 inches but should be treated in the same way. Like cilantro, it grows and goes over quickly. Be prepared to replace it through summer if you want a continuous supply. Cilantro, *Coriandrum sativum*, grows best in light well-drained soil. Give it some shade in the heat of the day but leave it in full sun to ripen the seeds.

If you have a corner that is shady for part of the day, slot in pots of chives, *Allium schoenoprasum*; parsley, *Petroselinum crispum*; and garden mint, *Mentha spicata*. Chives need plenty of water and feeding. If you let them die down in winter they will bounce back next spring. Parsley with its bright green curly leaves is as ornamental as it is useful. It has a long tap root so it needs a deep container, rich soil, plenty of water and feed. Mint is a spreader so it is best in a pot. It needs little attention apart from watering. There are numerous types of mint including spearmint, eau de Cologne, ginger, basil and pineapple—to name a few.

French tarragon, *Artemisia dracunculus*, has more flavor than the Russian, *A. d. dracunculoides*. They are not called dracunculus, or "little dragon," without reason. The roots are invasive and they are best grown in a container. They hate having wet feet and damp in general. Find a warm dry spot, water plentifully in the heat of the day for fast evaporation and add grit in the compost so that it drains through quickly. Don't overfeed or you will spoil the flavor.

Basil needs at least six hours sunshine a day, plenty of water and free drainage. Keep picking round the edge to keep it bushy and to prevent it from flowering. The ruffled plum-black variety, *Ocimum basilicum* Purple Ruffles, is one of the few hybridized herbs to be an All American Selections (AAS) winner. Basil is reputed to repel flies, so it's a good plant to have around the kitchen.

A neat way to make a tower of herbs, using alpine strawberries, thyme, parsley, and chives and catnip for a crown of blue flowers.

Oregano and marjoram are botanically the same and the names are often confused. The half-hardy sweet marjoram, *Oreganum majorana,* is the one most commonly grown in the herb garden. It has a more delicate taste than oregano, *Origanum vulgare,* which is the punchy herb most used in Italian, Greek and Mexican cooking. It needs long summers to develop the full flavor. Pot marjoram, *Oregano onites*, is a small easy shrub with a slightly bitter taste. They all like plenty of sun, sharp drainage and alkaline soil.

Some herbs, notably basil, sweet marjoram, and tarragon will not survive frost. Either move them into a frost-free place in winter or harvest them and start again next year.

Edible herbs for scent and color

Lemon verbena, *Aloysia triphylla*, is the best lemon-scented plant for culinary use. A half-hardy and ornamental deciduous shrub, it can grow up to 10 feet and has pale lavender flowers in late summer. It needs a large pot, staking, a warm sunny spot, fast draining compost, and plenty of water and fertilizer.

The half-hardy, shrubby pelargoniums from South Africa (not to be confused with their relations, the geraniums) have aromatic leaves and delicate flowers There are a whole range of flavors. *Pelargonium tomentosum* tastes of peppermint and has a trailing habit; *crispum* has a lemon aroma and grows to 12 inches, and the larger graveolens is rose-scented. There is even a pelargonium called Chocolate Peppermint.

For a splash of very bright color throw in some pot marigolds, *Calendula*, and nasturtiums, *Tropaeolum majus*. The flowers will brighten up salads. The tender pineapple sage, *Salvia elegans* Scarlet Pineapple, has striking red flowers in summer, grows to 36 inches and, while not considered the best for cooking, is worth having. It smells so deliciously of pineapple that every time you touch it you will be transported to a tropical island.

Growing herbs indoors

This takes a little more concentration due to lower light levels, fluctuating temperatures, central heating, and drafts. Put the plants by a sunny window, take out the leaders to stop them getting "leggy" as they reach for the light, and trim them to keep them bushy. If you can, alternate them with those outside. Many culinary herbs have dwarf versions which will fit neatly onto a windowsill.

Though herbs grown indoors need a little more attention than those grown outside, as they need extra trimming and turning to the light, even a couple of pots of culinary herbs on a bright window sill is attractive, and saves time and money.

The cosmetic garden

When making a cosmetic garden, don't simply think of lotions and tinctures. Aim for a garden which will be a health cure in itself by lifting the spirits and delighting the senses every time you enter it

Both for scent and beauty of form the old roses are unrivalled. They are the ones used by perfumers. Their flowering is so profuse and ravishing that they are forgiven by their many devotees for not flowering all summer long.

The earliest cultivated rose is the Apothecary's rose, *Rosa gallica* var. *officinalis*, a dark pink semi-double. The striped pink and white sport is *Rosa gallica* Versicolor or *Rosa mundi*, reputedly named after Henry II's mistress, Fair Rosamund. Both grow to about 90 cm/36 in. There are many beautiful varieties of the gallicas which have pure rose scent. Duc de Guiche is a cup-shaped, fully double rose in deep pink with a green eye. The flowers of Tuscany Superb are richest velvet, dark and plummy, set off with golden stamens. The damasks were brought to Europe from the East by the Crusaders in the fifteenth century. *Rosa x damascena* var. *semperflorens*, Autumn Damask or Quatre Saisons, has silky, crumpled, pale pink flowers and is a shrub growing to 120 cm/4 ft. The French perfumers favour it possibly because, unusually for an old rose, it has two flushes of flower. Elsewhere Professeur Emile Perrot is the most commonly used damask for making attar, or the essential oil, of roses. It's a soft pink rose, quite vigorous, growing to 150 cm/5 ft.

The *Rosa x albas* are exquisite both in form and sophisticated scent. *Alba Semiplena*, the White Rose of York is a vigorous shrub with large, cream-white, almost single flowers used for the production of attar of roses; Königen von

Dänemark is a true rose pink and considered one of the most beautiful in shape; Maiden's Blush is pale; Celeste is between the two with leaden foliage. They all grow between 150 and 180 cm/5 and 6 ft.

The moss roses *Rosa x centifolia* Muscosa are an enchanting group of roses with whiskers. They have a characteristic balsam scent. Hunslet Moss is a sturdy plant with deep pink flowers and a powerful scent; Cristata, formerly known as Chapeau de Napoleon, is particularly endearing as it has a moss formation like a cocked hat.

The Bourbons are Victorian roses. Louise Odier looks much like a pink camellia; Madame Isaac Pereire has deeper colour with a heady fragrance; Zephirine Drouhin is a climber to 2.7m/9 ft, and is particularly suited to cover a trellis above a seat as it is thornless. It is almost shocking pink but the scent is celestial.

Roses need sun and rich deep soil. If you buy them bare rooted in the winter cut them hard back, almost to the ground.

Lavenders associate well with roses and hide their bare

Above *Rosa gallica*, the romantic and ancient rose known over the centuries as the Apothecary's rose, and the Red Rose of Lancaster, was introduced to Europe from the East by the Crusaders in the 12th and 13th centuries.

Right The scented garden is a mirror image with repeat plantings.

THE ROSE BOWER – A SCENTED GARDEN FOR COSMETICS

Jasminum officinalis

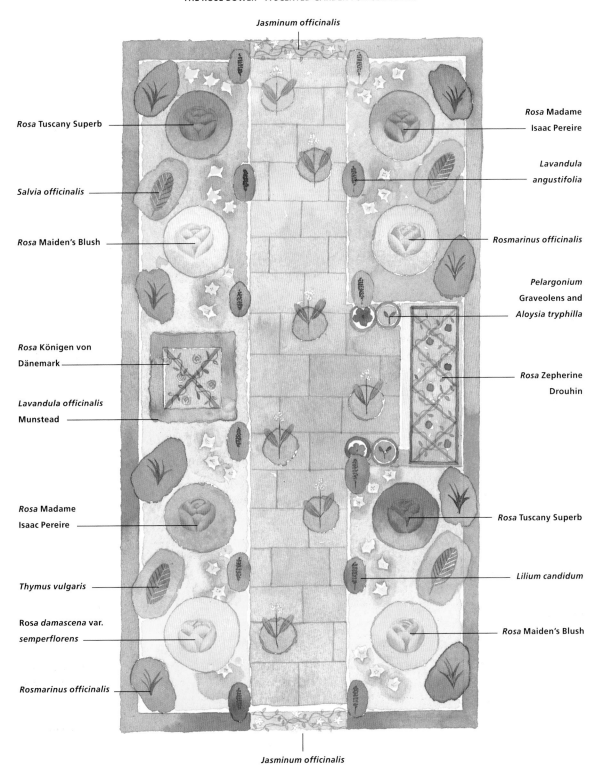

Rosa Tuscany Superb

Salvia officinalis

Rosa Maiden's Blush

Rosa Königen von
Dänemark

Lavandula officinalis
Munstead

Rosa Madame
Isaac Pereire

Thymus vulgaris

Rosa *damascena* var.
semperflorens

Rosmarinus officinalis

Rosa Madame
Isaac Pereire

Lavandula
angustifolia

Rosmarinus officinalis

Pelargonium
Graveolens and
Aloysia tryphilla

Rosa Zepherine
Drouhin

Rosa Tuscany Superb

Lilium candidum

Rosa Maiden's Blush

Jasminum officinalis

angustifolia is the one most used for lavender oil. It grows from 24 to 36 inches. Both the downy gray-green leaves and violet-blue flowers are highly scented. The Munstead variety is smaller at 12 inches with true lavender-colored flowers. Imperial Gem is medium-size with purple flowers. There is a pink variety Hidcote Pink. The white form, Nana Alba, is half-size and good for edging or pots.

Extracts of rosemary, *Rosmarinus officinalis*, are popular for cosmetic preparations. The flowers in late spring vary from pale to dark blue. There are many varieties. The Texas rosemary Arp has a slightly lemon scent. Pinkie is another American variety growing to 4 feet. The gilded-leaved Aureus will brighten up a dark background. The common sage, *Salvia officinalis*, is a good background player in the flower border and it is used as a fixative for perfumes. The oil is also extracted from the narrow-leaved, Spanish sage, *S. lavandulifolia*, which

has lavender flowers and balsam scent.

Lavender, rosemary and sage like a sheltered sunny spot and good drainage.

The common thyme, *Thymus vulgaris*, can be sown in cracks in the pavement in full sun for the extraction of aromatic thymol. A little shrub, it grows to 12 inches and has flowers ranging from white to purple.

The Madonna lily, *Lilium candidum*, symbol of purity in early Christian times, is not used any longer for cosmetics, however, the scent is exquisite and it would be a wonderful addition to the border. It is a magnificent plant growing up to 5 feet and producing some 20 white trumpet flowers in summer.

A rose bower, with *Rosa* sp. trained over the arbors, and the path edged with lavender. The path leads the eye, and throughout there is enchantment from the delicate colors and the fragrance.

The violet-scented orris root, used in talcum powder and perfumery, comes from the wild iris, *Iris* Florentina. It has sword-shaped leaves, is 24 inches high, which makes an upright contrast to the soft mounds of many herbaceous plants. The flowers, in late spring, are white with a violet tinge. Plant the tops of the rhizome roots slightly proud of the soil so that they can bake in the sun.

Let Roman chamomile, *Chamaemelum nobile*, used in beauty preparations, creep between roses. Only 6 inches tall, it will produce small apple-scented daisies all summer. Sow seed in spring outside, keep moist and protect from birds.

Find a little space for the scented, hardy, sweet violet, *Viola odorata*, the flower of Aphrodite, grown for perfume for 2,000 years. It only grows 6 inches tall with heart-shaped leaves. The flowers, which come in late spring or early summer, vary from white to violet. Given good rich soil and a little shade it will soon make a carpet. There is a pure white form, Alba.

Another striking plant for the cosmetic border is the aniseed fennel, *Foeniculum vulgare*. It has an airy grace. There is a bronze variety Purpureum. Yarrow, *Achillea millefolium*, growing to 12 inches or more, is a long-flowering border-worthy plant with flat white flowerheads like little dishes and feathery leaves.

Tender plants, which will not survive a frost but are very useful for cosmetics, are *Citrus bergamia*, the bergamot orange, from which orange flower water and bergamot oil are made. It will grow to 30 feet in the right situation. However, you can grow it in a large pot for many years, bringing it inside in winter. The rose-scented pelargonium, *Pelargonium graveolens*, which dates from the eighteenth century, is an upright sub-shrub growing up to 5 feet, with pale pink flowers each bearing two purple spots.

A good climber against a sunny wall or over an arch is jasmine, used since the sixteenth century for floral perfumes. *Jasminum officinalis*, the hardy deciduous jasmine, flowers most sweetly all summer long and grows to 30 feet if left unpruned. Also traditionally used for perfumery is the less vigorous but tender evergreen Spanish *J. grandiflorum* which has clusters of pink-tinged white flowers.

If there is room for a small tree, plant an elder, which can be used for elderflower water and skin lotions. The common European elder, *Sambucus nigra* (not to be confused with the American elder, *S. canadensis* which is poisonous) is a wild-

A summer border crowded with the heady scent of roses, *Nicotiana* and lilies.

looking hedgerow plant but there are some pretty variants. There is *S. nigra* Aurea with golden leaves; the fern-leaved variety *nigra* f. *laciniata* and prettiest of all Guincho Purple which has bronze leaves and pink-stemmed flowers. They all reach 20 feet at maturity and are both hardy and trouble-free.

With so much scent around you, do not let it escape. Trap it by enclosing the garden, keep it sheltered from winds and make it a place of peace and tranquillity. Face south to catch the sun and, if you like to sit in the dying rays of a summer evening, put in a seat facing west. Put an arch over the seat and clothe it in sweet-scented jasmine or a climbing rose. Plant the bracing aromatic rosemaries and lavenders where you can brush against them to make the garden into a place to delight the senses throughout summer.

The medicinal garden

Growing a few plants to make simple tisanes and cures for minor complaints is both practical and economical. Keep each type of plant separate in the way of the old monastery gardens to avoid confusion.

Medicinal herbs need to be treated with great respect. Some are highly poisonous. It is easy to overdose on even the most innocuous-sounding herbs. Only use the specified part of the plant–leaf, stem, flower or root–in the recommended dosage.

With this in mind however it is both useful and fun to have a patch of mild cures and basic domestic remedies to hand. In the fashion of the old monastery "physic" gardens, each herb is isolated from the next and clearly labeled to avoid confusion.

An herbal ladder for home cures

One way to make a garden of useful herbal cures would be a little row of low "simples" planted between the rungs of an old wooden ladder along a sunny path or by the house wall. More practical, as it would give you more space, would be a larger "ladder" made of bricks or wood.

Lemon balm, *Melissa officinalis*, is considered to be one of the elixirs of life, clearing the mind and lifting the spirits. It is believed to calm upset stomachs and relieve stress. A scented perennial growing to 36 inches high, it likes fertile soil, full sun, or dappled shade. Cut back after flowering for a fresh supply of leaves.

A soother and inducer of sleep is *Chamaemelum nobile*, the Roman chamomile. It grows to about 18 inches and has little daisy flowers throughout summer. There is a version with pretty double flowers, Flore pleno.

The taste and smell of peppermint, *Mentha piperita*, has been enjoyed for thousands of years worldwide. It is a gentle and effective digestive, good for many stomach problems. A hot tisane stimulates the circulation, reduces fevers and helps to clear colds. A hardy perennial, it grows up to 36 inches with little pink flowers in summer. It is a rampant spreader so plant it within a bottomless bucket to keep it in check.

Lady's mantle, *Alchemilla mollis*, is a neat plant with pleated lime green leaves. All summer long lacy yellow flowers (which are very useful for flower arranging) appear. When the plant looks tired cut it right back for a second flush. It is used in skin lotions for rashes and cuts, also as a gargle and is an anti-inflammatory agent.

The evergreen shrub, rosemary, *Rosmarinus officinalis*– one of the most useful all-round herbs–is said to be excellent for headaches and migraine. A hot tisane will help with colds, catarrh, sore throats and chest infections. The prostrate rosemary grows no more than 12 inches tall and twice the width.

Sage, *Salvia officinalis*, another all-rounder, is antiseptic. An infusion can be used as a gargle or mouthwash and the leaves can be chewed as a mouth freshener. A small variety is *S. officinalis* Kew Gold which only grows to 12 inches.

Above An infusion of sage, seen growing here with parsley and chives, makes an effective antiseptic.

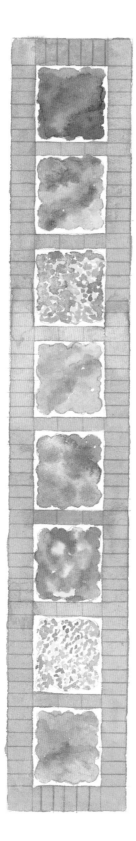

An herbal ladder

1 *Rosmarinus officinalis*
2 *Mentha x piperita*
3 *Alchemilla mollis*
4 *Sempervivum*
5 *Athemis nobilis*
6 *Melissa officinalis*
7 *Tanancetum parthenium*
8 *Salvia officinalis*

Minor burns and stings are quickly soothed with the leaves of freshly picked houseleeks, *Sempervivum*. Their decorative rosette leaves were prized by the Romans. They need little attention or water.

Feverfew, *Tanacetum parthenium*, is good for stings and bruises. It grows to 24 inches and has clusters of white daisies in summer. Cut back after flowering to keep a neat shape, since it is quite invasive.

Lemon balm, apple mint, rosemary, lavender and *Santolina* blend harmoniously in a herb border.

Cartwheel herb garden

A traditional way to grow herbs is in an old cartwheel with the spokes keeping each medicinal herb in its own compartment. As cartwheels are hard to come by these days, borrow the principle and build a bed in a cartwheel shape with bricks.

If there is room put a small tree in the middle. Witch hazel, *Hamamelis virginiana*, common in southeastern USA, and with a long history of use by Native Americans and settlers, provides the well-known cure for sprains and bruises. It is a beautiful, small (up to 15 feet), slow-growing, deciduous tree with clusters of yellow flowers for two months in the fall. If space is at a premium a sundial or a tall urn would make an alternative central feature.

If you make the wheel large enough you can go to town with some dramatic, tall medicinal herbs in the inner circle of "spokes"–the bright daisies of cone flower, the architectural angelica, the spiky milk thistle, comfrey with its watercolor flowers, and the phosphorescent evening primrose. The cone flower, *Echinacea*, is another herb that was used by Native Americans for treating wounds, and is one of nine species of colorful hardy perennials. *E. purpurea,* the purple cone flower, will reach 4 feet and has honey-scented, purple daisy flowers with spiky centers from midsummer to early fall. It grows in ordinary garden soil in sun. There are quite a few cultivars to choose from including the pure white *E. purpurea* White Swan.

For real drama, plant mullein, *Verbascum thapsus*, a giant hardy biennial which makes a rosette in the first year and in the second grows up to 6 feet tall, with large soft gray-green leaves and huge spikes of golden flowers through summer. Used by the Ancient Greeks to protect against infections and the Native Americans for painful joints, it has soothing antiseptic properties. Another fine tall garden plant is the hollyhock, *Alcea rosea*, a hardy perennial with rounded leaves and flowers like hibiscus. It can grow up to 6 feet or more.

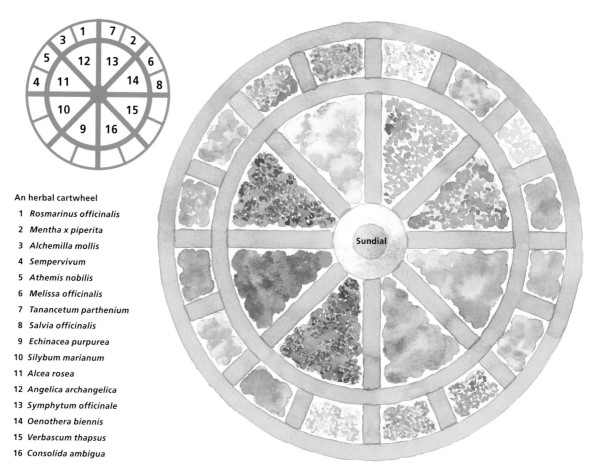

An herbal cartwheel

1 *Rosmarinus officinalis*
2 *Mentha x piperita*
3 *Alchemilla mollis*
4 *Sempervivum*
5 *Athemis nobilis*
6 *Melissa officinalis*
7 *Tanancetum parthenium*
8 *Salvia officinalis*
9 *Echinacea purpurea*
10 *Silybum marianum*
11 *Alcea rosea*
12 *Angelica archangelica*
13 *Symphytum officinale*
14 *Oenothera biennis*
15 *Verbascum thapsus*
16 *Consolida ambigua*

The flowers were used for relieving skin irritation and it has diuretic and soothing properties. There are many different varieties from near black through to pinks, reds and yellow. They like well-drained soil in sun.

Larkspur, *Consolida ambigua*, a healer of wounds, is an airy ornamental with violet-blue flowers in spikes all summer. The plants grow to 3 feet . A hardy annual, good for picking, it likes sun and fertile, well-drained soil. Note that the seeds are poisonous. Golden rod, *Solidago virgaurea*, commonly known as "woundwort," has antifungal and antiseptic properties used to promote healing. It is an upright perennial growing to 30 inches with numerous spikes of golden flowers from mid-summer to fall.

The angel's herb, angelica, *Angelica archangelica*, traditionally used for digestive problems, makes a dramatic statement in a garden. A robust hardy biennial, it can shoot up to 8 feet, with frothy lime green flowerheads. Best in rich soil in sun or part shade, it is grown from seed.

The common name for comfrey, *Symphytum officinale*, is "knit bone" and it has been used for centuries to help set broken bones. The leaves can be used to reduce inflammation (blanch them first in boiling water). It is a very useful plant to have in the garden for compost and to make liquid feed. One disadvantage is that it grows a large taproot and so it is difficult to move—it also has a tendency to spread. With long, hairy, lance-shaped leaves and bell flowers in watery blues, pinks, and white, it will flower all through summer. A damp situation is best but it is not fussy.

Evening primrose, *Oenothera biennis*, is a North American native. It has astringent and sedative properties and was used against whooping cough and asthma. It grows up to 4 feet and almost as much across with long oval leaves. Plant it in a well-drained, sunny position.

For spiky contrast try the milk thistle, *Silybum marianum*, traditionally used for liver complaints. It is a robust biennial foliage plant growing to 5 feet with white-veined, crinkled leaves and purple or pink thistle flowers.

In the outer circle plant the low-growing herbs as described for the herb ladder.

Checkerboard herb garden

If you have a dull area of paving, a neat way to grow medicinal herbs separately is to lift out some stones to make a pattern, then either plant the lower medicinal herbs in the gaps (removing the sand and putting in fresh compost) or dig pot depth holes and sink in containers of herbs. The advantage of this arrangement is that you can move them around and put evergreen types in during winter. For a bright splash throw in some *Tagetes patula*, the French marigold. It's a calming herb for digestion. It grows to 9 inches but flowers merrily red, yellow and gold all summer.

Checkerboard herb garden
1 *Hamamelis virginiana*   2 *Mentha x piperita*
3 *Alchemilla mollis* 4 *Rosmarinus officinalis*
5 *Tanancetum parthenium* 6 *Melissa officinalis*
7 *Salvia officinalis* 8 *Tagetes patula*
9 *Chamaemelum nobile*

Mixed borders

The shapes, textures and leaf and flower colors of herbs are invaluable in a mixed border. They will join the chorus line and can occasionally even take the star role.

When planning a border, look for foliage plants which will set off the flowers in leaf shape and color. Put soft and rounded plants next to spiky ones. The contrast will make both more interesting. Aim for at least 40 percent evergreens for winter and summer structure because the more ephemeral flowering plants come and go with the seasons. Pick up colors in flowers by repeating them in the leaves of the plant next to them. Alternatively, find contrasts to intensify the effect. One way to do this is walk around the border with some leaves and flowers to find happy combinations with the existing plants.

Most herbs are low key (though there are a few exceptions) and make excellent companion plants. Silver foliage is magical with pastels and creams or white. They show up well against dark evergreens like box or yew. Bronze and purple add fire to hot schemes in the red spectrum. Sharp limes and mellow golds will bring light into a dark garden.

Purple-blues

Many herbs have flowers in the blue range and combined with silvers work well with other gentle colors. One of the most useful for the border is catmint, *Nepeta cataria*. It makes grayish clumps topped with spikelike, lavender-blue flowers lasting many weeks, adding grace and profusion if dotted along the front of a border between other plants. If cut down it will do a repeat performance.

The medicinal hardy cranesbills, *Geranium* spp., will perform in the same way making soft rounded clouds along the border. There are many blue forms. *Geranium* Johnson's Blue will produce saucer-shaped, lilac-blue flowers all summer, growing up to about 30 inches and *G. magnificum* produces a glorious burst of violet flowers from midsummer onward.

Pulmonaria officinalis, lungwort, makes a cool and refreshing carpet in a semi-shady place. It is a perennial with spotted green and white leaves. The soft funnel flowers come up pink and then turn blue. Cambridge Blue has eggshell blue flowers emerging from pink buds. *Ajuga reptans*, bugle, is a tougher ground cover. It is evergreen and will survive in most conditions, spreading and making rosettes of oval leaves and spires of blue flowers.

Silvers

Many Mediterranean herbs have fine silver foliage which has evolved to reflect the sun. Quite a few are hardy evergreens which add structure to the winter border. They are worth having for their foliage alone. If their mustard-colored flowers clash with your color scheme they can be sheared off. They like a well-drained sunny position.

A study in painterly planting. The violet hue of French lavender is echoed in leaves of purple sage and the allium flowers and intensified by the fresh greens of angelica and lady's mantle.

The Artemisia family is almost entirely silver and comes in many leaf shapes, some are elegantly filigreed. *Artemisia absinthium*, wormwood, is a semi-hardy perennial growing no taller than 36 inches and a little wider in spread. A smaller variety is Lambrook Silver, with very silvery, deeply dissected leaves. *A. stelleriana* Boughton Silver is a compact low form with a broad hairy leaf in ghostly white-silver. Southernwood or lad's love, *A. arbrotanum*, is an erect shrub with a little more green in the leaves. The shrubby *Artemisia* Powis Castle has a feathery look and makes a billowing clump.

Cotton lavender, *Santolina chamaecyparissus,* has woolly whitish linear leaves and makes a little bush up to 24 inches. *Tanacetum argenteum*, a member of the tansy family, has frilly white-silver leaves, is even smaller, and grows to 8 inches.

Sage, lavender, and rosemary are invaluable evergreen ornamentals for setting off a mixed border and giving it a timeless air. The leaves are a gentle grayish-green and they all have blue flowers in summer, though you can find varieties in pink or white. The white flowering versions, *Salvia officinalis* Albiflora*, Lavandula stoechas* f. *leucantha* and *Rosmarinus officinalis* var. *albiflorus* have white flowers and would look stunning in a white and silver garden.

Old fashioned pinks, *Dianthus* spp. have a sweet clove scent used for potpourri. They form compact, evergreen, silvery mounds reminiscent of cottage gardens. There are so many beautiful forms: Mrs. Sinkins is a charming raggedy white.

To add some upright spikiness to contrast with soft and round forms, try the silver sea holly, *Eryngium maritimum*. It grows wild on the coastlines of Europe and is happy in well-drained soil in a sunny site. The foliage is truly metallic blue-silver and the powder-blue flowers are good for drying. Whiter forms for the border are the elegant Miss Wilmott's Ghost which reaches 36 inches and has steel-blue flowers. Most spectral of all is *E. giganteum* Silver Ghost.

A plant for big impact is great mullein, *Verbascum thapsus*, a biennial with woolly gray-green leaves which reaches up to 6 feet 6 inches, crowned with great spires of yellow flowers. The globe artichoke, *Cynara cardunculus* Scolymus group, a culinary and medicinal herb, is a good border plant for a touch of architecture. Very erect, it has deep cut silvery leaves up to 30 inches long and produces big thistle flower heads.

Eucalyptus, the gum tree, does not need to grow into a giant. If you stool it (prune it to near base) each spring it will stay as a shrub and grow large, glaucous blue-silver leaves. It is a useful foliage plant for flower arranging.

Limes and golds

To bring sunlight into your garden, go for gold. Many of the most useful herbs have a golden form. There is a golden sage, *Salvia officinalis* Icterina, which has yellow variegations. Kew Gold is a compact form, 12 inches, with golden leaves; *Origanum vulgare* Aureum is golden marjoram; *Melissa officinalis* Allgold is golden lemon balm; *Tanacetum parthenium* Aureum is golden feverfew, and *Thymus x citriodorus* Argenteus is golden thyme.

The yellow-leaved bay tree, *Laurus nobilis* Aurea can be kept compact if clipped. A sunny shrub or small tree is *Sambucus nigra* Aurea, the golden elder. A highly resilient plant which will grow in almost any conditions, it can take hard pruning if you want to keep it as a shrub. The leaves emerge golden and mature to lime green. A particularly beautiful variety, though not for medical use, is *Sambucus racemosa* Plumosa Aurea which has finely cut leaves and an airy appearance.

A plant that is invaluable through summer is the herbaceous *Alchemilla mollis*, lady's mantle. In spring the neatly pleated light green leaves are followed by clouds of fresh green flowers on long stems, the whole effect being of a splash of lime.

A lime-green climber to brighten up a wall is the golden hop, *Humulus lupulus* Aureus. A deciduous climber with toothed leaves, it can grow to 20 feet in a season. It will happily twine itself up vertical strings without any tying in. They can be cut right off to ground level in the fall to emerge anew the following spring.

Many herbs have yellow flowers. Particularly dramatic among them is the architectural Jerusalem sage, *Phlomis fruticosa*. An upright plant with woolly grayish leaves on a stout stem growing to 36 inches, it has whorls of hooded yellow flowers in summer. Angelica, *Angelica archangelica*, is another statuesque pale green plant with bold large acid green flowerheads.

Angelica archangelica, with its umbels of lime green flowers in early summer makes a dramatic and architectural statement in a relaxed cottage garden.

Orange-reds

If you have a fiery border of reds and oranges, purple and bronze foliage plants will intensify it. For a low soft mound try purple sage, *Salvia purpurea* Raspberry Royal or, for some airy loftiness, the bronze fennel, *Foeniculum vulgare* Purpureum. The bronzy bugles make good groundcover. *Ajuga reptans* Atropurpurea has purple-brown leaves with a high satin sheen. Burgundy Glow is an eye-catcher with some bronze, some green, and some raspberry-colored leaves. Bugles like moist soil in sun or dappled shade.

The eau de Cologne mint, *Mentha x piperita citrata,* has leaves which appear reddish purple in sun or bronze in shade and it produces purple flowers in summer. It is a spreader so keep it under control by planting it within a bottomless bucket. Tender purple basil, *Ocimum basilicum* Purple Ruffles, (which tastes just as good as the green) comes with crinkled leaves, while Dark Opal has near black leaves.

Herbs with purple flowers include the stately hollyhocks, *Alcea rosea*. The poisonous foxgloves, *Digitalis purpurea*, and monkshood, *Aconitum napellus* (once used for poisoned

A herbal show stopper, bergamot, which flowers continuously from mid-summer to fall.

arrows and with legal restrictions in some countries) are nonetheless great border plants. Monkshood with its strange, hooded indigo flowers in late summer, grows up to 5 feet. It likes deep moist soil in shade.

While the flowers of most herbs are discreet, bergamot, *Monarda didyma*, is an exception. A hardy perennial growing to about 36 inches, with hairy lance-shaped leaves, it has unusual feathery flowers at the end of the stems.

The cone flower, *Echinacea purpurea*, is pretty showy. It flowers from midsummer to fall making a clump up to 5 feet with big daisy-type flowers which are good for picking. Leuchstern, a smaller variety, has purple-red flowers.

A red flowering herb not to be missed for the hot border is *Lobelia cardinalis*, the cardinal flower. Both leaves and stems have a bronze tinge and the flowers are a true scarlet with purple bracts. It is an upright perennial growing to 36 inches and needs reasonably moist soil in sun or dappled shade.

Walls

A few herbs, most notably wallflowers, toadflax, wall germander, and houseleeks thrive in the dry poor conditions found in cracks in walls to make decorative vertical interest.

If you have an old brick or stone wall, a good use of vertical space is to grow some herbs in it. An obvious candidate is the spice-scented wallflower, *Erysimum cheiri*. Some of the new varieties will flower from late winter through to summer. A particularly handsome variety is the vigorous *Erysimum* Bowles Mauve. It looks wonderful close to blue-mauve wisterias, although it does needs sun, so ensure it is not covered or shaded by it.

Wall germander, *Teucrium chamaedrys*, is an evergreen, shrubby perennial with shiny oak-like leaves and tubular rosy-purple flowers both in summer and fall. The alpine toadflax, *Linaria alpina*, has long-lived spires of violet snapdragon flowers which occasionally come out in soft yellows, pinks or white. There is an purple variety, Purpurea, and a pink one, Rosea.

Valerian, *Centranthus ruber* (not to be confused with the medicinal valerian *Valeriana officinalis*) grows wild in cliffs and chalk pits. It will live happily in a wall, flowering throughout summer on 12-inch stems and seeding for the next. It is a sturdy perennial with numerous pink and occasionally red or white flower clusters.

The houseleek, *Sempervivum*, makes little rosettes of succulent leaves, useful for plucking and rubbing on stings. It has little star-shaped flowers in summer. There are many varieties that can make an enchanting collection. Though not to be trodden on, they will live happily in paths as well and are engaging planted on the roof of an old shed.

The ideal wall will be situated in full sunshine, and have lime mortar rather than concrete. If there are no little crannies and you cannot loosen the mortar, drill out some holes. Fill with free-draining chalky compost with a few seeds in it or put in a little plantlet. Keep moist until it establishes . Providing there are other crannies, it should self-seed and need no further upkeep.

The blowsy pink or white form of valerian self-seeds and colonizes merrily in a sunny wall needing no attention.

Herbs underfoot

Thymes and chamomile will make pools of foliage and flower in gaps between paving stones. They can survive a limited amount of treading on, releasing their delectable scent to passers-by.

The best group of herbs for paving are the mat-forming thymes. They have fragrant foliage and flowers and are favorite bee plants. Given plenty of sunshine and fast drainage, they are quite tough and can survive some treading on which will release their aromatic scent.

There are many to choose from but perhaps the best are the true carpeters growing only a few inches high, with up to a 36-inch spread. *Thymus serpyllum* Annie Hall has purple flowers; Pink Chintz has gray leaves and pink flowers; Rainbow Falls has gold variegated leaves with mauve flowers.

When planting thymes in paving excavate a hole with enough room for the root run and put in some grit and sandy compost for fast drainage. Do not let it dry out too much until the plant has established.

Apple-scented chamomile survives even more treading than the thymes. It can be used in paving in the same way, or it lends itself to the scented lawn or seat. The best variety for the purpose is non-flowering, bright green *Chamaemelum* Treneague. A chamomile lawn is a labor of love. It needs careful and constant weeding until it knits together, hopefully in the second season. However it is bliss to stretch out on in the summer and once achieved needs little aftercare, staying emerald green in the hottest summers while the grass goes brown. As *Chamaemelum* Treneague does not flower it cannot be sown from seed. However, unlike the Roman chamomile, *C. nobile*, it grows more sideways than up, will concentrate on leaf growth, and needs less cutting and fussing over.

To make the lawn, weed the area well and remove any stones. Rake it flat and roll it. Plant the little chamomiles about 6 inches apart. Don't walk on it except to weed it until the following year. After that upkeep consists of clipping over in the fall.

Gravel gardens

Many people find gravel an excellent solution for low maintenance in their gardens. Herbs are extremely well suited to sun-drenched positions as they imitate conditions in the wild with a cool root run under stones and rocks. There is no grass to mow and once established, the herbs will need little or no watering. If you cover the area with weed-suppressant matting followed by a good layer of gravel there will be no weeding either. Make criss-cross slits in the matting to plant the herbs. Any of the Mediterranean plants will thrive.

Above Thymus doerfleri Bressingham makes a lush carpet of magenta flowers in summer along with houseleeks and lavender in a gravel garden.

Right The evergreen thymes planted between gaps in harsh paving slabs soften the appearance. The gravel around the plants acts as a water retaining mulch and keeps down weeds.

Planting herbs

Choose healthy and reliable plants, give them a flying start with good soil or compost, the right sun or shade aspect, water and feed. Avoid putting them under stress and they will reward you by performing well.

Buying plants

When buying, choose bushy plants of good shape and check around the leaves for signs of pests and disease. A well-grown plant will show some root but should not be pot-bound. If there is a mass of roots growing out of the bottom it will need repotting immediately.

The potted culinary herbs which can be found in supermarkets are usually grown under highly artificial conditions and are likely to keel over at the first whiff of real life. They usually come in the tiniest and flimsiest plastic pots. However, they can be grown on with a little care. Repot them and harden them off carefully. Take them from windowsill to a sunny spot outside (bringing them in at night), gradually giving them more exposure until the plant shows signs of new growth and loses its soft sappy look.

Potting up

The first task when you bring any young plant home is to replant it. Choose a container with drainage holes and with room for growth of root and shoot. Put a good layer of crocks (pieces of broken clay pots) over the drainage holes to prevent them getting clogged, followed by a layer of grit. Herbs like fast drainage.

For container growing it is best to buy potting compost which is designed and balanced for the plant's needs and will also be sterile. The broad choice is between soil-based and soil-less composts. If the pots are to stay outside it is probably best

to use a soil-based type. The main ingredient is sterile topsoil which holds both nutrients and water well. The soil-less types drain fast which means more feeding and watering. They dry out extremely quickly and, once they have done so, they do not take up water easily. When it comes to weight considerations on roofs or for hanging baskets, the soil-less composts win because they are lighter.

Soil outside

Fertile living soil is the single most important factor for plant growth. Good soil contains air and water, vegetable and microscopic animal remains and is vibrant with algae, fungi and bacteria which in turn feed the plants. One sign of good soil is a healthy worm population, another is a vibrant weed population. Even though most herbs will survive in poor soils they will be happiest in free-draining neutral to alkaline loam.

Which type of soil? Soils fall roughly into six groups. Loam is the ideal. It is crumbly, pleasant to handle and it drains well while holding nutrients and water. Sand is gritty, free-draining but low in nutrients. Chalk is poor soil, pale and stony. Peat is dark brown and spongy and can become waterlogged. Silt is silky to the touch and packs down like clay, a moldable, sticky

Right Chives, cotton lavender, and the long-flowering cone flower are planted in large clumps for visual impact.

Soil testing

Chalky soils are very free draining (some too much so) and clay soils are heavy and slow draining. A simple way to get a general indication of how free draining your soil is to put a few tablespoons of soil in a large jam jar with a lid, three quarters fill with water, shake, and leave it to settle overnight.

Slow-draining clay   Average draining loam   Fast draining sand

cold soil but rich in nutrients. Peat, silt, and clay are bad drainers and not ideal for herbs. If your soil is very chalky or full of clay the addition of well-rotted compost will help. Clay soils will also benefit from the addition of sharp grit. If your soil is really slow draining the solution is to raise the beds.

Acid or alkaline Depending upon the amount of of lime in your soil it will be either acid or alkaline. To find out which, buy an inexpensive kit from a nursery or garden store. The test works on a pH test of one to fourteen. Seven is neutral, anything above is alkaline and anything below is acid. Most herbs like neutral to alkaline soil. If your soil is acid, add some hydrated (builder's) lime about two weeks before planting. Manure will make the soil more acid but never apply it at the same time as lime as they react against each other.

Improving drainage The easiest way to get round soggy soil is to make raised beds. They only need to be 12 inches high to work. Once you have made the edging from bricks or timber break up the existing soil and fill with good loamy topsoil. Another method is to double-dig and put a layer of coarse grit mixed with top soil on the lower layer (sub-soil) for drainage. Incorporate plenty of bulky well-rotted compost on a regular basis. It is the magic improver for every type of soil.

Compost

Well-rotted compost is full of micro-organisms which bring life to soil. It opens up compacted soils allowing vital oxygen to the roots. The worms will return and aerate it further. Compost improves both drainage and water-holding capacity. If possible, have a compost heap, compost bin or a worm bin.

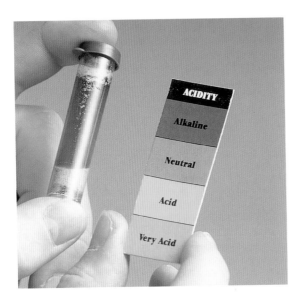

Using a proprietory soil test kit, compare the color of the solution with the accompanying color chart to determine the pH of your soil. This soil is alkaline.

Incorporate the compost, once it has really rotted down on an annual basis or more often.

If you don't have the time or the room for composting, there are various commercial products that you can buy. Seaweed is an excellent soil conditioner. It binds soil particles together and improves drainage. Spent mushroom compost– a mixture of peat, manure and chalk–is alkaline which suits herbs, and is a good soil conditioner. Well-rotted horse manure is an excellent soil improver and fertilizer. It can be too rich however for those herbs which are used to a meager diet. They tend to lose potency in favor of lush leaf growth.

Propagating herbs

Buying plants can be expensive and it is both satisfying and easy to propagate for those who have the time and inclination. While seeds need daily attention until they are ready for planting, cuttings, once done, only need to be kept from drying out and dividing plants brings instant gratification.

Growing from seed

As you start to build up a collection of herbs, consider growing from seed. There is more choice, and it is much less expensive and very gratifying. Some herbs spring so easily from seed that very little effort is needed. Lady's mantle, angelica, marigold, dill, feverfew or fennel usually seed themselves if left to their own devices. All you need to do then is to thin out the seedlings when they appear in spring.

Some herbs need to be sown from seed where they are intended to grow. Parsley, chervil, cilantro, borage, and dill will "bolt," or run to seed, if they are transplanted.

Outdoors Prepare the soil by weeding thoroughly. Rake it over to get rid of all the lumps and stones. The aim is to make a fine texture, or "tilth," so that the seedlings can push through effortlessly. Trample it down to make a firm surface by walking slowly to and fro. With a few exceptions (check the packet), seeds are sown in spring when both soil and weather are warm. You can hasten this process by covering the area you want to plant with polythene for a couple of weeks before sowing.

Make a "drill," or little channel. Water the soil so that it is pleasantly moist. Always sow as thinly as possible. If the seed is very fine (some are like dust) mix it with a little gardener's (silver) sand. Cover to about twice the depth of the seed. With fine seed, sprinkle a little shop-bought compost over the top. To hasten germination cover with a cloche. You can improvise one out of an old plastic water bottle sawn off at the bottom and with the cap left off, or make a mini-polytunnel with hoops of wire covered with polythene. Don't forget to label the spot. If you want a more free and random planting, scatter the seed and mark the area with chalk or sand.

Prepare the ground by hoeing to a fine tilth

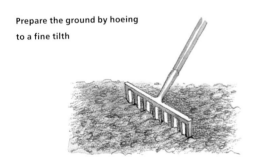

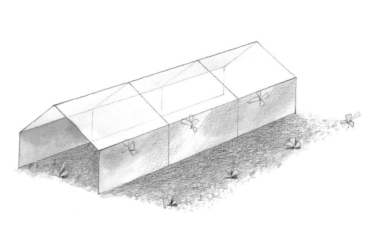

Cover seeds with cloches to protect them

SEED SOWING

Sowing seed comes in particularly handy for annuals and to keep up a continual supply of those useful culinary herbs which need to be replaced through summer.

1. Three-quarters fill a seed tray (or flower pot) with proprietory compost which will have been sterilized. Level the surface with a block of wood.

2. Either water gently from using a fine rose or wait until the seeds are sown and place in a tray of water and let the compost soak it up by capillary action.

3. Sow seed thinly following the directions on the packet. If the seed is very fine mix it with some sand. Sift compost over the seeds to a depth specified on the packet. A depth of two-and-a-half times the size of the seed is a rough guide.

4. Label! It is easy to confuse seedlings.

5. Cover with glass and newspaper or black polythene. Check every day for germination.

6. As soon as you see signs of life, take off the cover and put in a light warm place out of direct sunlight.

Indoors The advantages of sowing seed indoors are that you can start earlier which is particularly useful for tender plants in a cool climate, and although there is more work attached, you do have more control.

Use clean pots or seed trays and sterile seed compost. Firm it and water well. When you have sown the seed, cover with black polythene or glass with newspaper on top to cut out the light. Put it in a warm place. Check every day and when the seedlings appear, take off the coverings and put the tray in a light place but out of direct sunlight.

When they are large enough to handle, pot them on in potting compost. Always pick up seedlings by the leaves, not the stem which is easily broken. Accustom them to life outside gradually, protecting them from cold and bringing them in at night until they establish.

Dividing plants

An easy way to multiply many herbaceous perennials is to divide them. Catmint, bergamot, comfrey, marjoram, and yarrow will split easily to make more plants. Cut down the leaves and dig them up carefully in the dormant season. Tease the roots apart, making sure that each section has some root and shoot. Sometimes you will find a dead-looking center. Cut this out and you will give the plant a new lease of life. If the roots are large and entangled, sink two garden forks back to back through the plants and into the ground and lever them apart. Sometimes you need to cut through tough sections with a sharp knife. Replant the divisions. Chives and many other bulbs make little bulblets. These are easy to split up.

Increase your stock by dividing herbaceous perennials. Plunge garden forks (or spade and fork) back to back right through the plant and into the ground.

Lever and tease apart to make new plants.

When pricking out use a dibber and touch the delicate seedlings as little as possible. Pick the plants up by the leaves not the stem.

Cuttings

Cuttings are the most effective way to make more plants from shrubby herbs that would take a while to grow from seed. Rosemary, lavender, sage, hyssop, bay, and southernwood will root from "greenwood" cuttings.

In mid- to late summer, pull off a young healthy shoot without a flower taking with it a "heel" (a tiny bit of the woody stem). Trim the heel to make a clean cut and remove the lower leaves with a sharp knife. Dip it into hormone rooting powder though herbs usually manage well without it.

Plant in moist, free-draining compost (half sand/half compost) plunging them almost–but not quite–to the level of the few remaining leaves on top. Keep the cuttings in a cool shady place until you see signs of growth. Spray the leaves with mist every day for the first week at least or cover with a cloche or polythene. The cover must not touch the leaves and it needs to be taken off fairly frequently to avoid dripping condensation. Never let them get too wet as they might "damp off." They generally root within a month.

Maintaining herbs

The secret of good maintenance is little and often. Keep a watchful eye on your plants to make sure that they have everything they need to keep them in top health.

Watering

Once your herbs are planted in fertile garden soil they should need little day-to-day attention, but very young herbs and moisture-lovers (like angelica and mint) appreciate watering in hot weather. In the garden it is better to give plants a good drenching less often (encouraging the roots to go down) rather than frequent sprinkling. Plants in containers are more demanding. In summer they may need a daily drench and hanging baskets may need watering twice on hot days.

The best time to water is in the cool of evening when evaporation rates are at their lowest.

Small sticks placed around the pot prevent the polythene from touching the cuttings while they root.

Feeding

As a general rule through summer, give herbs in containers a liquid feed once a week and once every two weeks for those you want for cooking, cosmetic or medicinal use. If they start to look too lush and sappy (which will attract pests) cut down the amount. Do not feed in winter.

Liquid seaweed is a good proprietary feed that is rich in trace elements. As an alternative, slow-release fertilizers in granular form can be sprinkled onto the compost.

Weeding

Weeds and even grass take a surprising amount of nutrients out of the soil so it is worth keeping on top of them for this reason alone. Never let them seed. As the old saying goes, one year's seed means seven years' weed. A good mulch of about 3 inches of shredded or chipped bark laid on top of the soil is a highly effective way of keeping the weeds at bay. The added bonus is that it cuts down the watering and improves soil structure. Lay it on when the soil is wet after a good rainfall or watering. Other good mulches are mushroom compost, garden compost and leaf mold.

Deadheading encourages more flowering and prolongs the season. Always cut just above a joint or node to prevent die back.

Deadheading

To encourage a longer flowering season deadhead the flowers that will repeat, such as nasturtiums, cone flowers and chamomile. This will stop them putting their energy into making seed and they will produce more flowers. Also deadhead the spreaders that self-seed a little too freely.

Pests and disease

You can do much to avoid problems by observation and being quick off the mark. Keep plants in top health by taking note of their particular needs of sun or shade, shelter, soil, and feed. Avoid putting plants under stress with lack of water, competition with weeds or poor air circulation. Don't let them become pot-bound.

Pick up problems early. Much can be achieved by removing any affected leaves at the first sign of trouble. Once any infection is spotted, clean hands and tools to avoid spreading it. Either burn the infected material or tie it up in a trash can liner and dispose of it.

Inspect new stock to make sure that it does not carry disease and, if given a plant which might be ailing, separate it.

Provide camouflage by planting a mixture of different types of plant so that they are less of a target for pests who find food by sight. Encourage biodiversity by providing food and habitats for useful predators.

Herbs as trouble shooters

Herbs have not been in existence for thousands of years without acquiring defenses, including spikes, prickles, hair, stings, pest repellents, insecticides and pungent, confusing aromas. Their flowers are designed to attract flying insects which will eat their enemies. With all this armor, they protect the plants around them.

Many have the type of simple flower that will bring in friendly predators. Some herbs produce a pest repellent. Southernwood, *Artemisia abrotanum*, was traditionally tied up in sachets to repel moths and fleas. *Artemisia vulgaris* was known by the Anglo-Saxons as "midge plant" because of its deterring properties. *Tagetes minuta*, a relative of the French marigold (not to be confused with pot marigold, *Calendula officinalis*) excretes sulfur compounds called thiophenes from its roots which inhibit eelworms from attacking it or the plants around it. Its common name is Stinking Roger.

The onion family, particularly garlic, *Allium sativum*,
excretes a strong sulfur smell which is unattractive and confusing to pests. The Mediterranean shrubs, lavender, rosemary and cotton lavender, surround themselves in a cloud of powerful essential oils that few pests would choose to cross.

Herbal pesticides

One way to take advantage of those herbs with insect repellent qualities is to make an infusion to spray on other plants. Pour boiling water onto a few handfuls of leaves and allow to soak overnight. Strain next day and add a squeeze of washing-up liquid. This will help it to adhere to leaves. Spray on a calm day in the evening after the good insects have retired for the night. It won't keep, so use it up within a few days.

Herbs as good neighbors

Companion planting has been practiced for centuries, but it is only now that serious scientific research is being carried out on the interaction between plants and the chemical reactions that take place, often at root level. Garlic is said to help prevent blackspot on roses and to improve the scent; yarrow, to help oil production and vigor in other plants; chervil to make

Marigolds with bright daisy-type flowers, attract lacewings and hoverflies, both friendly predators in keeping down aphids.

radishes hotter. There is much that is still to be discovered about the complex chemical makeup of plants and the effects that one plant has on another.

Pruning and trimming

Pruning encourages fresh growth as well as keeping plants in good shape. Many herbs have rampant habits and need tidying up from time to time for the sake of appearances. Some profit from being chopped to base. You can prolong the life of others with a timely haircut to prevent them from going to seed. The shrubby Mediterranean herbs get leggy and woody if not kept in trim.

Good practice is to consider the entire plant from all angles before making a cut. Look at the overall shape, check to see if there are rubbing or crossing branches or any sign of disease. Plan what you are going to do. Always use sharp tools that are up to the task in hand.

Shrubby herbs In warm countries, pruning can be done after flowering. Trim lightly after flowering to stop the plant putting its energy into making seed. It will then have time to restore itself before winter. Pruning will stimulate tender new growth which will be knocked back by frost. In colder climates, leave any major surgery until the weather is set fair in spring.

Thyme, rosemary, and lavender do not put out fresh growth easily once you have cut into the old wood. Thyme will not take hard pruning but appreciates a light clipping over after flowering. Prevent lavender from going woody by trimming off the new season's growth in late spring to within a few inches of the previous year's. Let young rosemaries grow freely. Cut off any lanky or damaged growth in late spring. If you have an old woody specimen of lavender or rosemary it is worth a last ditch attempt to rejuvenate it by cutting back half the shoots to half the length and doing the other half the following year.

Old sage plants which have a tendency to sprawl generally recover from being cut almost right down to base. Avoid them getting to this state by pruning off the tips on young plants to encourage them to branch and become bushy. Cut back close to the framework of old wood in spring. Trim off the flowers when they fade.

Cotton lavender, curry plant, and the silver-leaved *Artemisias* (wormwood, southernwood, mugwort, and tarragon) positively prosper from being cut to base occasionally. You will help to maintain the beautiful foliage if you cut off the flowers before they bloom. In spring, cut back the shrubs to about half the size. Every two or three years, or when they start to look untidy, rejuvenate them by cutting down to a few inches from the ground.

By midsummer many of the culinary herbs, mint, lemon balm, and marjoram, will have become coarse and unappetizing, so cut them hard back to base for a fresh new crop. The cranesbills, catmint, comfrey, and lady's mantle can be treated in the same way. Chop them right down when they start to look tired and they will spring up again for a second performance.

Roses To keep roses in good health, it is important to remove dead, damaged, and diseased wood and also branches that cross or rub against each other. The aim of formative pruning on a young shrub rose is to make an open goblet shape for air circulation and balance. To avoid "die back" always make a clean-angled cut just above a bud. Choose a bud which is pointing in the direction that you want the new shoot to grow. Some of the old roses put out a lot of "suckers" which should be cut off at ground level. Don't deadhead the once-flowering, old-fashioned roses or you will miss out on the hips. Tidy up in the fall, shortening long shoots that might be damaged by winter winds, and renovate in spring.

Pinch pruning is a technique of constantly nipping out the shoot tips to produce bushy plants. As the plant does not have the chance to make flowers it will divert its energy into making side shoots and can be shaped into a neat ball shape, pillar or cone.

If you need a little height in your garden quickly, you can grow a fast climber like a hop or honeysuckle into a tall "standard" (lollipop or mushroom shape) within a single season. You need an upright pole securely fixed into the ground with a strong circle of wire firmly attached to the top. After planting, cut off all but a couple of strong shoots. When they reach the top let them branch out over the circle of wire and clip them into shape. They will soon make a sculptural feature.

Harvesting

Timing is all. Harvesting herbs is a continual process depending on the plant and how it will be used. Only pick perfect flowers as they are opening, taking care not to damage them. Harvest in the morning of a dry, sunny day when they are at their most powerful in flavor and scent. The essential oils dissipate by the end of the day. Wait until the dew has gone so that they are as dry as possible.

Culinary leaves are picked throughout the year but harvested at their peak just before they come into flower. Pods and seed heads are gathered when ripe and going brown but before they disperse. Roots are at their most potent at the end of summer or in the fall.

Herbs for drying

Shake out any insects and wipe the leaves clean if necessary. Do not wash the herbs (except roots) as you want them to dry quickly. The faster they do so the better they will be. Sun drying is the traditional method in hot countries, though this does fade them. In cooler regions, find a dry, warm, well-ventilated spot in a shed or in the kitchen. If you can provide heat so much the better. The ideal temperature is 75 to 80°F.

Small-leaved herbs and flowers on long stems, like lavender, can be hung up in bunches. Bay leaves and other large-leaved herbs are best picked off the stem and laid out on a drying tray. You can devise one of these out of an old oven tray covered in netting or brown paper with holes punched in it for air circulation. Rose petals are treated in the same way. Roots should be washed and sliced. They can take more heat and can be dried in an oven on a cool setting of about 130°F.

The herbs will be ready for storing when they are dry and crackly. The next step is to get them away from light and air as soon as possible as they deteriorate quickly. Store them in airtight jars in a dark place.

Preserving

Use fresh herbs to make herb oils. Make sure that the oil completely covers them. Steep them for a week. Strain them off and repeat until the flavor is to your taste. Herb vinegars are made in the same way although they take up the flavor more quickly. A stalk of a dried herb is a pretty addition to vinegar and will last indefinitely. Unlike the vinegars, oils with garlic do not keep and pose a health risk of botulis. Make them fresh and use immediately, and discard any leftover quantity.

Freezing is an excellent way to keep all culinary herbs to hand, particularly basil, parsley, chives, and mint which lose their flavor when dried. Chop them up and freeze with a little water in an ice cube tray. Transfer them into plastic bags for storing in the freezer. You can make long cold drinks look supremely elegant if you add ice cubes containing a single perfect herb leaf.

Right Herbal oils and vinegars add the taste of sunny Mediterranean cuisine to dishes throughout the year and make fine presents.

Lemon + Lime Vinegar

Herb + Garlic Oil

French Tarragon Vinegar

Herb directory

THE HERB DIRECTORY IS A SELECTION OF 55 HERBS ARRANGED alphabetically by botanical Latin name. The common name is also given but common names can often lead to confusion because many plants have different common names in different parts of the world. For example *Calendula officinalis* (the botanical Latin name) is known as pot marigold in North America, but simply as marigold elsewhere; featherfew is also known as feverfew but its botanical name is universal, *Chrysanthemum parthenium*.

    The Directory provides specific information on the derivation of each herb's name, both botanical and common, and its native origins and habitat. There are gardening tips and information on a particular herb's attributes and its value for cosmetic, culinary, and medicinal purposes. The chapters that follow look in more detail at the most popular and important herbs for those applications. ◼

the herb directory

- *Achillea millefolium* Yarrow
- *Alchemilla vulgaris* Lady's mantle
- *Allium sativum* Garlic
- *Allium schoenoprasum* Chives
- *Alo barbadensis* Aloe vera
- *Aloysia triphylla* Lemon verbena
- *Althaea officinalis* Marshmallow
- *Anethum graveolens* Dill
- *Angelica archangelica* Angelica
- *Anthemis nobilis* Chamomile
- *Anthriscus cerefolium* Chervil
- *Arnica montana* Arnica
- *Artemisia abrotanum* Southernwood
- *Artemisia dracunculus* Tarragon
- *Borago officinalis* Borage
- *Calendula officinalis* Pot marigold
- *Carum carvi* Caraway
- *Chrysanthemum parthenium* Feverfew
- *Coriandrum sativum* Cilantro
- *Crataegus laevigata* Hawthorn
- *Echinacea* Cone flower
- *Foeniculum vulgare* Fennel
- *Gingko biloba* Maidenhair tree
- *Glycyrrhiza glabra* Licorice
- *Hypericum perforatum* St John's wort
- *Hyssopus officinalis* Hyssop
- *Iris germanica* Orris root
- *Jasminum officinalis* Jasmine

- *Laurus nobilis* Bay
- *Lavandula angustifolia* Lavender
- *Melissa officinalis* Lemon balm
- *Mentha* Mint
- *Monarda didyma* Bergamot
- *Myrrhis odorata* Sweet cicely
- *Nepeta cataria* Catmint
- *Oenothera biennis* Evening primrose
- *Ocimum basilicum* Sweet basil
- *Origanum vulgare* Oregano
- *Origanum majorana* Marjoram
- *Panax ginseng* Ginseng
- *Pelargonium* Sweet-leaved geranium
- *Petroselinum crispum* Parsley
- *Piper methysticum* Kava kava
- *Rosa* Rose
- *Rosmarinus officinalis* Rosemary
- *Salvia officinalis* Sage
- *Salvia sclarea* Clary
- *Sambucus nigra* Elder
- *Serenoa repens* Saw palmetto
- *Silybum marianum* Milk thistle
- *Symphytum officinale* Comfrey
- *Thymus vulgaris* Thyme
- *Trigonella foenum-graecum* Fenugreek
- *Valeriana officinalis* Valerian
- *Zingiber officinale* Ginger

yarrow *Achillea millefolium*

Common yarrow, a hardy perennial and native to Europe, grows as a rampant weed in fields and hedgerows, where it can vary from a low, creeping form to a tough plant up to 24 inches high. It has flat heads of minute five-petaled white flowers. A. millefolium v. rosea and A. filipendulina respectively have pink-cream or bright yellow flowers.

■ **History** The plant's generic name is believed to derive from the Greek hero Achilles, who is said to have used it to heal his soldiers' wounds during the Trojan War. Accordingly, it has been called *herba militaris*, the military herb, knight's milfoil, bloodwort, and staunchweed.

■ **Characteristics** The leaves are about 4 inches long, dark green, downy and feathery, and the stems are pale green, rough, and angular. The plant flowers from early summer to late autumn.

■ **Growing tips** The plant thrives in full sun but will tolerate shade. It is easily grown in any type of soil, even in poor soil, and is increased by dividing the roots.

■ **How to use** Fresh leaves may be used in salads. They have a slightly pungent taste and are very aromatic.

The leaves and pounded flowers are frequently used in healing infusions and cleansers, particularly for problem skins. Combined with other herbs it is a valuable addition to steam treatments and face packs for deep cleansing and a valuable scalp treatment for itchy and oily hair.

Medicinally, the dried aerial parts of the plant are used, including the flowers. Yarrow is a good antiseptic for urinary infections and can be helpful in cases of diarrhoea. The herb can also stimulate the appetite.

■ **How to take** Take 15 to 20 drops of liquid tincture twice daily.

lady's mantle *Alchemilla vulgaris*

*A delightful and very useful border plant with its pleated leaves and lime-colored flowers that look wonderful in flower arrangements. In medieval times it was thought to have magical as well as medicinal properties. **A. mollis** is a native of the mountainous areas of Europe, Asia, and America. It will tolerate a wide range of conditions.*

■ **History** The name derives from the Arabic *Alchemych* which means "little magical one"—so called because of the way that the leaves hold perfect drops of rain and dew. The word *mollis* means "softly hairy." In medieval times it was reputed to preserve youth and was prescribed for female complaints. The pleated leaves are rather like a little cloak and it became associated with the Virgin Mary from whence came "Our Lady's mantle."

■ **Characteristics** Lady's mantle is a hardy, clump-forming perennial. The leaves are soft, downy, rounded, and toothed. They are sea-green to lime. The yellow to lime flowers are very small but numerous and carried on branching stems. Height 24 inches. Spread 30 inches.

■ **Growing tips** Buy small plants, which will soon multiply, or sow the very fine seed in seed trays in the fall or spring and plant out after dangers of frost. If you cut the plants back hard after flowering, there will be a resurgence of fresh green leaves and another flush of flowers.

■ **How to use** Use the astringent juice from the leaves as a tonic for oily skin or

add an infusion to healing creams for dry skin. Used as a toner, an infusion soothes inflamed skin caused by minor infections and sun and wind burn. It also has mildly bleaching properties while the leaves, soaked in warm water and laid on the face, will help reduce wrinkles and fine lines.

Medicinally, this herb has been used in the treatment of menopausal disorders for centuries. Its actions are due to the herb's natural astringent properties that help to control irregular bleeding. Taken internally, lady's mantle can help regulate excessive or irregular periods, and applied externally, it is useful for the treatment of vaginal discharge.

■ **How to take** Take 20 drops of liquid extract twice a day.

garlic *Allium sativum*

A member of the onion family, garlic is an indispensable flavoring in cooking and is widely used throughout Europe, the Middle East, the Far East, Africa, the West Indies, Mexico, and North and South America. A native of Asia, it is widely cultivated in warm climates, but in cooler conditions never reaches its maximum flavor potential.

History Garlic has been used medicinally and as a flavoring for at least 5,000 years, and has been cultivated in the Mediterranean region since the time of the ancient Egyptians. The Anglo-Saxons grew it too, and gave it its name: *gar*, a lance, and *leac*, a leek.

Characteristics The straight, rigid stem, topped by a spherical pink or white flowerhead, grows to a height of 24 inches. Each bulb is made up of several cloves, which may have white, pink, or purple skin, encased in a paper-like sheath. The size, number, and flavor of the cloves vary according to the variety and the climate.

Growing tips Garlic grows best in well-drained soil in a sunny position. Cloves are planted in autumn or early spring to mature in summer. They should be planted 1 inch deep and up to 8 inches apart, and should be given a good start with the application of a general fertilizer.

How to use Garlic complements the flavor of meat, fish, vegetables, salad dressings, sauces, and egg dishes. The raw juice of garlic is a cosmetic aid, and it has extraordinary antiseptic and healing properties. Unguents containing equal quantities of garlic, beeswax, and honey are reputed to cure baldness.

Medicinally, garlic oil helps keep the lungs clear of infections. The treatment of pneumonia, bronchitis, and asthma should be followed up by a preventative dose of garlic daily. The risk of heart disease, due to cholesterol deposits, can be reduced by taking a regular dose of garlic. Garlic has powerful anti-microbial activity and can be applied directly to an infected area. Fungal infections, often difficult to control, can be helped by a garlic application. New research is suggesting that garlic contains anti-cancer substances, but this is still a new area of study.

How to take Take 2 to 3 garlic capsules daily, with a meal, or 1 to 2 teaspoons of a liquid tincture daily. Do not use deodorized garlic preparations if you can avoid it.

chives *Allium schoenoprasum*

Chives, a member of the onion family and grown from bulbs, are native to northern Europe, where they may sometimes be found growing wild. They also thrive in temperate regions of North America. The leaves have a delicate, onion flavor and are widely used in cooking, particularly in egg and cheese dishes.

- **History** In the Middle Ages, chives were known as "rush-leek," from the Greek *schoinos*, rush, and *prason*, a leek. They were used in antiquity, and have been cultivated since the sixteenth century.

- **Characteristics** Chives grow in clumps, with their round, hollow, grass-like leaves reaching a height of 9 inches. Some varieties, *A. sibiricum* for example, may be 15 inches tall. The stems are firm, straight, smooth and, like the leaves, bright dark green. The flowers, which bloom for two months in midsummer, form round, deep mauve or pink heads and are attractive used as a garnish.

- **Growing tips** Chives flourish in a moist, well-drained soil and like partial shade, though they can tolerate full sun. Seed may be sewn in late spring or late summer, or the plants may be increased by dividing the clumps in mid-spring. Chives grow particularly well in pots, and are a good choice for a kitchen windowsill garden.

- **How to use** Snipped chives—for it is easier to cut them with scissors than chop them with a knife—give a hint of onion flavor in many dishes, from scrambled egg to cheese soufflé.

  Medicinally, chives are high in vitamin C and iron. For this reason they are considered to be a highly nutritious food and excellent for building up the blood.

  Chives have a mild stimulant effect on the appetite and can aid in digestion.

- **How to take** Simply eat a good-sized sprig of whole herb daily.

aloe vera *Alo barbadensis*

Aloe vera will only grow outside in warm countries but it makes a good spiky houseplant elsewhere, thriving on neglect. It has been an important medicinal herb since the Egyptians. Aloe vera comes from the Mediterranean, the Cape Verde Islands, and the Canary Islands. It has spread to the Caribbean and South America.

History The name comes from the Greek/Hebrew *allal*, meaning bitter. *Aloe* is its native name in South Africa and *vera* means true. The Egyptians used it for embalming. Jesus was wrapped in linen soaked in "myrrh and aloes" as "the manner of the Jews is to bury" (*St John* 19: 39–40). It is said that Aristotle prized it so much that he asked Alexander the Great to conquer Socotra off the coast of Arabia—the only place where it was known to grow at the time. It was introduced to Europe in the tenth century.

Characteristics Aloe vera is a tender, clump-forming perennial. It makes a sprawling rosette with grayish/olive-green pointed, succulent leaves with creamy teeth along the margins. Yellow bell-shaped flowers emerge in panicles like a candelabra throughout the year. Height 24 to 36 inches. There are over 300 species of *Alo*.

Growing tips Aloe vera needs to be kept above 50°F to survive. In cooler countries it will be perfectly happy in a cool greenhouse or conservatory. Propagate by taking an offshoot from the base. Let it dry for a day and pot it up into free draining soil mixed with sharp sand. Leave it to establish in a warm place. Take off the basal shoots in summer to keep the parent plant in trim and to harvest it. Feed with liquid fertilizer in spring. Let it dry out between waterings. Do not water in the winter. In hot countries, keep it outside in sun or part shade. (It is subject to legal restrictions in some countries.)

How to use Medicinally, the gel has good moisturizing properties when applied to dry skins, especially in cases of eczema. The gel can also be used to treat minor burns. When taken internally the same gel can ease indigestion and inflamed bowels although care should be taken by those with sensitive stomachs since it can have laxative actions.

How to take Twice daily, one tablespoon of juice or apply to the skin as required.

Warning Internal use not advisable during pregnancy.

lemon verbena *Aloysia triphylla*

One of the most delightful of scented plants, lemon verbena has a strong citrus aroma that is at its most powerful in the early evening. A native of South America, it thrives best in hot climates, where it will grow up to 5 feet tall and almost as wide. It is, therefore, a good choice as a back-of-the-bed plant in sunny borders.

■ **History** The plant was brought to Europe by the Spaniards, and was used as a source of fragrant oil for perfume.

■ **Characteristics** The woody stems are tough, angular, and have many branches, giving the plant, which grows to a height of 5 feet, its bushy, spreading characteristic. The long, pointed-oval and pale green leaves are about 4 inches long and ½ inch wide. The flowers, which grow in clusters along the stem, are pale purple and bloom in late summer.

■ **Growing tips** A perennial, deciduous shrub, lemon verbena is hardy to a temperature of 40°F. This means that in cool climates it needs protection in winter. It may be grown from seed, or from soft cuttings taken in July and grown in sandy soil under cover. The bush should be pruned in spring to contain its growth and remove dead wood.

■ **How to use** The strong citrus aroma is used to flavor stuffings for meat, poultry, and fish, in fish dishes and sauces, fruit salads, poached fruit, soft drinks, and cream sweets. The herb may be substituted for lemongrass in south-east Asian dishes.

Lemon verbena produces a sharp, lemon-scented essential oil used in cosmetics for oily skins and used extensively in the perfume industry. It should be used with caution in skin preparations as it may cause irritation. The flowers and leaves of the herb are not particularly pungent but can be used in bath bags and potpourri.

Medicinally, a tea made from the dried leaves can help reduce nausea and flatulence, ease nervous anxiety, and help calm palpitations.

■ **How to take** Drink 2 to 3 cups of lemon verbena tea daily.

marshmallow *Althaea officinalis*

A member of the hollyhock family, marshmallow has small but attractive flowers carried without stems. It is grown throughout Europe, in Australia, Asia, and eastern North America. The mucilage, which comprises about 30 percent of the roots, stems, and leaves, was used to make the confection known as marshmallow.

■ **History** The plant's medicinal properties have been recognized since ancient times. Mallow features in a second-century BC herbal, and was illustrated in another from the sixth century AD.

■ **Characteristics** The plant grows to a height up to 4 feet, with a spread of 18 inches. The long, tapering root is cream colored and fleshy, somewhat resembling a parsnip, and the bright green leaves are heart-shaped and irregularly toothed, with pronounced veins in a yellowish green and a downy coating on both sides. The five-petaled flowers are saucer-shaped, white or pink. They bloom in late summer, but with no fragrance.

■ **Growing tips** Marshmallow likes damp conditions, and may be found growing wild on marshland. A perennial, it can be grown from seed sown outdoors in spring, or increased by dividing the roots in the fall. It is important to keep the plants moist, especially during a dry summer.

■ **How to use** The young leaves and shoots may be shredded and added to salads and soups; the roots may be parboiled, then fried in butter.

An infusion made from the cleaned, chopped roots of marshmallow is a mucilageanous lotion suitable for tired eyes when applied on lint pads. An infusion of the leaves and flowers is softening, gentle, and antiseptic and is used in creams and lotions for skin conditions. It is also an effective hair conditioner to counteract hair loss.

Marshmallow has been a favorite herb for the treatment of colds and chest infections. Its soothing action can be helpful to inflammations of the gut and lower intestine, especially in conditions like colitis. Combined with licorice, it can make a very effective anti-ulcer remedy. Conditions such as asthma and bronchitis have been reported to respond well to the herb.

■ **How to take** 2 to 3 tablets (100 mg) dried extract after meals.

dill *Anethum graveolens*

The plant originates from southern Europe and western Asia, and its use is recorded far back in time. As with so many umbellifers, this hardy annual yields two separate culinary components, its seeds and its feathery leaves, which are, somewhat ambiguously, known as dill weed.

■ **History** The herb was used medicinally by doctors in both ancient Egypt and Rome; indeed it was the Romans who introduced it to northern Europe. After centuries in obscurity, it surfaced again in medieval times, when its use was widespread, particularly in Scandinavian countries like Norway and Sweden.

■ **Characteristics** Dill grows to a height of 30 inches, with a spread of 12 inches. The green stem, which is hollow and smooth, branches out at the top and carries large flat umbellifers of bright yellow flowers that bloom in midsummer. The leaves are ultra-fine, feathery, and dark green, and have a taste similar to that of parsley. The flat, oval seeds are parchment-colored, and have a rather bitter flavor.

■ **Growing tips** The plant is easy to grow from seed sown in late spring or early summer. It favors poor, well-drained soil and a sunny position and, since it does not transplant

well, should be sown where it is to grow. If the plant is sited close to fennel, cross pollination is possible.

■ **How to use** The fresh leaves are used in salads, fish dishes, and sauces to serve with fish.

Medicinally, dill is very effective for the relief of colic and pain associated with trapped wind. A water made from dill seeds is still readily available from pharmacies and generations of children have been helped by its gentle calming action on the gut.

■ **How to take** Taken as dill water (gripe water) 1 to 2 teaspoons is normally sufficient to release trapped wind.

angelica *Angelica archangelica*

A statuesque plant (up to 8 feet) in fresh light green, and it looks great in a border. It is used for flavoring liqueurs. The stalks are candied and the leaves are eaten as a vegetable in Scandinavia. A. atropurpurea, is similar in appearance while Dang gui or Chinese angelica, A. polymorpha var. sinensis, is one of the foremost herbs in Chinese medicine.

■ **History** It is said that the name *Angelica archangelica* may have come from the Greek *angelos* or messenger. It became known as the "angel herb" as it was thought to protect against evil as well as having great healing powers.

■ **Characteristics** Angelica is a hardy biennial or short-lived perennial with upright stems and pinnately divided leaves. Pale green flowers are produced in umbels in late summer followed by elliptic notched seeds. It is attractive to beneficial insects.

■ **Growing tips** *A. archangelica* likes rich moist soil in sun or semi-shade. Propagate by seed in autumn or spring. It will self-seed freely. If you are growing it for medicinal properties of the root, cut off the flower before it forms. Dang gui, *Angelica polymorpha v. sinensis*, is grown as a commercial crop.

■ **How to use** The sweet and musky essential oil of angelica is used extensively in perfumes, toilet waters, and colognes and can be added to soaps and bath oils. As it is relaxing and fragrant a few drops added to the bath water or foot bath is to be recommended.

Medicinally, angelica appears to have a beneficial effect on the circulation of blood and body fluids. For the treatment of menstrual cramps and fluid retention there can be no better herb to take than angelica. It can also be used as a remedy for stomach upsets, gastric ulcers, and migraine sickness.

■ **How to take** 20 drops of liquid tincture made from the stems, roots, and leaves can be taken 2 to 3 times a day or try a capsule containing 200 mg of the dried herb daily.

■ **Warning** Avoid in pregnancy. Angelica may disrupt blood pressure in large doses. Some people may suffer light sensitivity due to the content of a substance called furocoumarins contained in the angelica plant. This may cause skin irritations.

chamomile *Anthemis nobilis*

Apple-scented chamomile, a perennial plant of the composite family, is one of the daintiest of herbs. A low-growing type known as Roman chamomile can be grown as an effective ground cover to form a green and white daisy-flowered lawn.

■ **History** It is said that the ancient Egyptians used chamomile as a cure for ague. Its use was widespread in the Middle Ages, not only in southern Europe, where it originates, but throughout northern Europe, too. It is mentioned as a medicinal herb in both John Gerard's and Nicholas Culpeper's herbals.

■ **Characteristics** The plant may grow to a height of 12 inches. It has shallow, fibrous roots and a green, hairy, branching stem. The leaves are finely cut and feathery, and the flowers, are compact and creamy white with yellow conical centers.

■ **Growing tips** The plant is easily grown from the division of runners, which are planted out in early spring, and from seed; it is also a prolific self-seeder. It prefers a fertile, moist soil in a sunny position, but will cling tenaciously to life in a poor, well-drained soil. There is a non-flowering variety, Treneague, which some people prefer to use for lawns.

■ **How to use** What the plant lacks in culinary uses—there are none—it makes up for in other ways. Infusions made from chamomile are mild enough to be used on dry and normal skins and are so gently

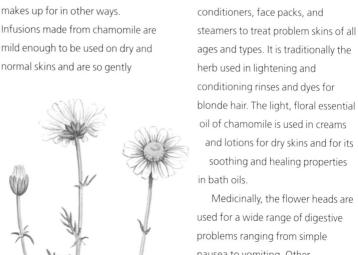

astringent and healing they can also be used in cleansers and conditioners, face packs, and steamers to treat problem skins of all ages and types. It is traditionally the herb used in lightening and conditioning rinses and dyes for blonde hair. The light, floral essential oil of chamomile is used in creams and lotions for dry skins and for its soothing and healing properties in bath oils.

Medicinally, the flower heads are used for a wide range of digestive problems ranging from simple nausea to vomiting. Other abdominal pains such as menstrual cramps benefit from this herb due to its ability to reduce spasm in the smooth muscle that surrounds the internal organs.

■ **How to take** Try taking 15 to 20 drops of liquid tincture twice daily, alternatively drink 2 to 3 cups of chamomile tea daily.

chervil *Anthriscus cerefolium*

*Chervil, together with chives, parsley, and tarragon, is one of the **fines herbes** mixture used in French cooking. It is also one of the herbs used in **ravigote** sauces, and is often blended with tarragon to flavor béchamel and other creamy sauces. It is a hardy annual, one that is easy to grow but that quickly goes to seed.*

- **History** The plant is a native of the Middle East, southern Russia, and the Caucasus, and was almost certainly introduced to northern Europe by the Romans. It became one of the classic culinary herbs.

- **Characteristics** A member of the umbellifer family, chervil is closely related to parsley. It grows to a height of 20 inches with a spread of about 8 inches. It has flat, light green, and lacy leaves, which have a slightly aniseed-like aroma and turn reddish-brown as the plant matures. It blooms in midsummer, producing flat umbellifers of tiny white flowers.

- **Growing tips** The plant can easily be grown from seed planted in early spring or late summer in the position where it is to grow; a trough or a window box is ideal. A succession of sowings will produce a harvest well into the winter. It likes a moist, shady position, and should be kept well watered.

- **How to use** The leaves quickly lose their flavor and are best added fresh to a dish just before serving. They can be chopped into softened butter to serve with broiled meats or poultry; added as an aromatic garnish to creamy soups; and stirred into egg and cheese dishes.

Use the leaves to make astringent infusions, or extract their juice and use as a tonic wash which is particularly beneficial for sallow skin. Add the infusion to cleansers and conditioners to soften fine lines and wrinkles.

Medicinally, the leaves are used before the plant comes into flower for conditions such as indigestion and a hot chervil poultice can be used for joint pains. However, chervil is not used much in medicinal situations any more.

arnica *Arnica montana*

Arnica is an attractive plant for the border or rock garden as its bright yellow, daisylike flowers keep coming all summer. It has been used as a medicinal herb since biblical times. Arnica montana can be found growing on the sub-alpine areas of Europe, Canada and North America. As it is becoming rare in the wild, it is protected in some countries.

History The name comes from the ancient Greek, possibly from *arnakis* or lamb's skin as the soft texture of the leaves is rather like it, or from *ptarmikos*, or sneezing, as it has this effect. Arnica is also commonly known as "mountain tobacco" as the leaves and roots were smoked. Pier Andrea Mattioli, physician to Emperor Ferdinand I, recommended it in his tome of herbal medicines *Commentarii* in 1544. It was much favored in Austria and Germany from the sixteenth century. It is said that Goethe (1749–1832) took it in his old age for angina. The American Indians used it to treat muscular pain, sprains, and bruises.

Characteristics A fully hardy rhizomatous perennial growing to up to 24 inches high and spreading to 6 inches. It has a basal rosette of ovate, light green, hairy leaves 2 to 7 inches long. It bears golden-yellow, scented, daisy-type flowers about 2 inches across throughout the summer.

Growing tips Being an alpine, arnica needs a cool climate, sandy soil rich in humus, sharp drainage, and a sunny position. It can take two years to germinate from seed. The easiest way to propagate it is to grow it from a piece of the rhizomatous root, which

divides easily. (There are legal restrictions on arnica in the USA.)

How to use Medicinally, arnica has remarkable properties and can assist in the healing process. Bruises, cuts, and abrasions all respond very well to arnica cream. Sports injuries, when caught early, will improve quickly with an arnica preparation.

Recent studies on arnica have suggested that internal use should be avoided. The United States has ruled the plant unsafe for internal use, however, short-term internal use, under the supervision of a qualified practitioner, may be helpful for the control of some heart conditions.

How to use Internal use should be restricted to professional guidance but homeopathic preparations containing arnica are considered to be very safe. Cream products can be safely applied to any part of the body.

Warning Avoid in pregnancy. Do not take internally unless under professional advice, an overdose can prove fatal!

southernwood *Artemisia abrotanum*

With the delightful popular names of "lad's love" and "old man," southernwood, a bushy shrub, is grown in many cottage gardens and herbaceous borders as a decorative and strongly aromatic plant, which is said to be repellent to bees. The French called it **garde-robe** *because they used it in wardrobes to ward off moths.*

■ **History** Dioscorides described the plant as having such fine leaves that it seemed to be "furnished with hair," while in his herbal Culpeper attributes it with the power of curing baldness. He recommended rubbing a paste made of the ashes and salad oil on the head or face to promote hair growth.

■ **Characteristics** The plant can grow to a height of 36 inches, with a spread of 24 inches. The woody stem has many soft, branching shoots covered with strong, feathery, gray-green leaves. The tiny flowers, which appear in late summer, are golden-yellow.

■ **Growing tips** The plant likes a rich soil and a sunny position. Take soft cuttings of new shoots in summer and cover until firmly rooted. Protect the plants in harsh winters, and prune them in late spring.

■ **How to use** The pungent leaves were used in Italy as a flavoring for meat and poultry forcings, and for cakes, but have few culinary uses now.

This most delicious-smelling herb prevents dandruff and promotes hair growth. To make a hair tonic to alleviate these conditions pour 5 tablespoons each of a strong Southernwood infusion and mild eau de Cologne into a bottle and

shake well. Use diluted in an equal quantity of warm water to massage into the scalp twice a week. This is a tonic particularly suited to oily hair. The dried plant keeps its perfume well in potpourri and is excellent for keeping insects away.

Medicinally, the whole plant is used to rid the body of worms and other parasites. It was once used as a powder or treacle to treat children with worms.

■ **How to take** Self treatment is not recommended since this herb may induce severe vomiting, vertigo, muscular cramps, and delirium.

tarragon *Artemisia dracunculus*

A distinction must be made between the true French tarragon or estragon *and its Russian counterpart, A.* dracunculoides, *which is much coarser, and has paler leaves and a bitter taste. Contrarily, the latter is easier to grow! French tarragon has a subtle flavor and is one of the four ingredients of the* **fines herbes** *mixture.*

- **History** The plant, a hardy perennial, originates from southern Europe. The reference in its name to a "little dragon" is thought to derive from its folkloric reputation of curing the bites of snakes, serpents, and the like.

- **Characteristics** The plant grows to a height of 36 inches, with a spread of up to 18 inches. The leaves are dark green, long slender, and pointed, about 3 inches long toward the base of the plant and considerably smaller at the tip of the stems. The flowers are lime green and formed in loose clusters. But the plant neither flowers nor sets seed in a cool climate.

- **Growing tips** You can grow French tarragon by planting a piece of the rhizome, complete with buds, in spring, or by taking cuttings of young shoots in summer and growing them under a cloche. The plant likes good, well-drained soil, and a sunny, sheltered position.

- **How to use** Tarragon has a strong and distinctive flavor, and must be used sparingly, especially as it is usually associated with delicate dishes such as chicken, white fish, creamy sauces, and egg and cheese recipes. Fresh sprigs of the herb are used to flavor vinegar for use in salad dressings and sauces.

  The wonderfully aniseed smell of the essential oil of tarragon is very soothing and when applied, a few drops in a base oil, it is an effective massage for easing pre-menstrual discomforts. It can also be used to perfume soaps and bath oils. The dried herb can be powdered and added to soap balls.

  Medicinally, the herb has been used in the past to treat toothache and to stimulate the appetite but its medicinal use has fallen off in modern times in favor of its culinary applications.

borage *Borago officinalis*

The bright blue, star-shaped flowers of borage make it one of the prettiest of herb plants, though the leaves, dark green, downy, and with no fragrance, are unremarkable. It is a hardy annual, a native of northern Europe, and grows well in the temperate regions of North America.

History Borage has, over the centuries, been accredited with legendary powers. Pliny called it *euphrosinum* because it was said to bring happiness and joy wherever it grew. The ancient Greeks and Romans looked to it for comfort and courage, and this belief in its capabilities was revived in the Middle Ages.

Characteristics The leaves have a flavor reminiscent of cucumber. The plant grows to a height of about 18 inches with a spread of 12 inches. It has an untidy, straggling habit, compensated for by the cloud of blue flowers that grow in arched clusters and persist throughout the summer months.

Growing tips Borage is easy to grow from seed sown outdoors in spring. It likes a sandy soil and sunny position, but will tolerate a heavier soil and partial shade. The plant self-sows seed freely and will, in this way, colonize a large area.

How to use Borage flowers and leaves are the traditional decoration for gin-based summer cocktails and may be set in ice cubes to garnish other drinks. The flowers may be used to garnish salads, and candied for cake decoration.

Medicinally, borage has mood-enhancing effects. The exact constituents of this plant have not been identified but its reputation for "lifting the spirits" dates back to 1597 when John Gerard included it in his book *The Herball, or Generall Historie of Plantes*. In this book borage was said to "drive away sorrow and increase the joy of the mind." During this time the leaves and flower were often made into wines and given to men and women to make them "glad and merry."

How to take Try a liquid tincture made from the leaves and flowers. A dose between 15 to 20 drops can be used twice a day.

Alternatively a capsule of borage oil can be taken at a dose of 500 mg daily.

pot marigold *Calendula officinalis*

Pot marigolds are hardy annuals that flower all summer and seed themselves for next. Yellow or orange, their bright daisy-like flowers attract insects, bees, and butterflies. The light green leaves have a pungent aromatic scent. If the flowers are open in the morning it is said that you can be sure it will be a fine day.

- **History** In *The Winter's Tale*, Shakespeare described their habit of closing up when there is no sunshine. "The Marigold that goes to bed wi' the sun, And with him rises weeping." Calendula comes from the Latin *calendulae* meaning the first day of the month, as they always seem to be in flower then. Originally from the Mediterranean and Iran, marigolds were used in India, ancient Greece, and Rome for medicinal purposes and as a dye.

- **Characteristics** Fast growing annuals with an erect, sometimes spreading, habit on branching stems. The soft hairy leaves are lanceolate. Height and spread 24 inches. The orange, yellow, or apricot flowers are daisy-like with ray florets about 3 inches across.

- **Growing tips** Marigolds must be among the easiest plants to grow. Almost any soil will do as long as it is not waterlogged and they like the sun best. Sow seeds outside in spring, cover with a little soil and they will pop up and colonize merrily. Deadhead to encourage even more flowers.

- **How to use** The juice or infusion of the leaves and flowers of this most healing of herbs are used, both fresh and dried, in a great variety of oils, creams, lotions, face packs, and steamers to remedy all skin problems from acne to sunburn, tired eyes to warts. Alone or with other herbs it is an effective colorant and conditioner. The green, fragrant essential oil of calendula has healing and rejuvenating properties when used in cosmetics, massage oils and baths. Used in a base oil, it will effectively reduce scarring.

  Medicinally, the entire flower head is used to help speed recovery from injury. Calendula promotes the healing process in nearly all body areas. It has been used to heal gastric ulcers, gall bladder disease, congested lymphatic nodes and inflamed skin. An external cream application can be very useful in cases of eczema and other skin inflammations.

- **How to take** Take 15 to 20 drops of liquid tincture twice daily or drink 2 to 3 cups of marigold tea daily.

caraway *Carum carvi*

Caraway is a two-in-one plant. The bright green, feathery leaves have a mild flavor, somewhere between that of parsley and dill, while the seeds, a spice, have a strong aroma and pungent taste. The plant is grown commercially for its seed in northern Europe, the United States, and North Africa.

■ **History** The plant was extensively used by the Romans and was well established in English kitchens in the Middle Ages, when it was cooked with fruit, especially spit-roast apples, and in cakes and bread. The leaves were chopped into soups and salads. In Germany and Austria—still the prime users of the plant—the seeds were cooked with vegetables, especially cabbage and its preserved form, *sauerkraut*.

■ **Characteristics** A biennial, the plant grows to a height of up to 24 inches, with a spread of 12 inches, and has thick, tapering roots rather like those of parsnip. The leaves resemble those of the carrot in shape. The flowers, in umbellifer clusters, are white tinged with pink and appear in midsummer. The pointed-oval seeds are dark brown, almost black.

■ **Growing tips** The seeds are sown outdoors in early fall. They like a good soil and partial shade. They should be harvested in the second year, just before they ripen. Hang them upside down to dry, the heads tied into a paper bag to catch the falling seeds.

■ **How to use** The leaves may be used in salads and soups, the seeds in baked goods, in dumplings, cream cheese, and meat dishes such as goulash.

Traditionally the seeds are used in potpourri and scented clothes bags to repel moths. When crushed they are very pungent but will give an unusual fragrance to soap and toilet waters. The essential oil is also used commercially to perfume soaps and toilet water. When chewed the seeds will sweeten the breath.

Medicinally, caraway is well known to reduce colic in babies and flatulence in adults. Its calming effects on the bowels are based in its anti-spasmodic activity on the bowel's muscular wall. Adding some caraway seeds to an herbal tea will help fight a cold or flu.

■ **How to take** Adults: 20 drops of liquid tincture taken twice daily.

Babies and children: 5 to 10 drops added to warm water. Give this between meals.

feverfew *Chrysanthemum parthenium*

With its bright lime green or yellow-green leaves that retain their color through the winter, feverfew is a year-round decorative garden plant. It is low growing, bushy, and vigorous, quickly thickening up, spreading and self-seeding. The pretty white flowers, which may be like single or double daisies, dry well for flower arranging.

■ **History** Its medicinal uses are well chronicled by Gerard who, in his *Herball*, said that the dried plant was useful for those "that are giddie in the head … melancholike and pensive," and by Culpeper, who recommended it for "all pains of the head coming of a cold cause."

■ **Characteristics** Various forms of the plant may grow to a height of anywhere from 9 inches to 24 inches. The deeply cut leaves are brightly colored, and have a sharp, unpleasantly bitter taste. The flowers, which are produced through the summer and into mid-fall, are thick and daisy-like, with yellow centers.

■ **Growing tips** The plant will thrive in the poorest of soils, even in paving cracks and walls. Ideally, it likes a well-drained soil and a sunny position. You can easily grow it from seed or by root division.

■ **How to use** The plant's bitter taste rules out culinary uses, but is worth tolerating for its medicinal and cosmetic properties.

Feverfew was one of the ingredients of the famous Gervase Markham's seventeenth-century skin tonic, which was reputed to purify and repair unsightly complexions. A lotion made with a handful of leaves and flowers simmered for 20 minutes in ½ pint of milk, cooled and strained, will nourish dry skin, remove pimples and blemishes, reduce soreness, and lighten freckles. Alternatively stand the herb in buttermilk for a week, strain and apply to the face, until dry.

Medicinally, feverfew contains many chemicals, one of which (*parthenolide*) has the ability to reduce inflammatory substances, associated with migraine headaches, being released from blood cells. Feverfew has also been used in other types of headache remedy as well as in the treatment of minor fevers, rheumatism, and arthritis.

■ **How to take** Try taking 20 drops of liquid extract twice daily, however if your migraine is sensitive to alcohol, use a capsule of the dried herb daily.

cilantro *Coriandrum sativum*

Both the green feathery leaves and the spherical seeds of cilantro are indispensable in the kitchen. Bunches of cilantro, which looks like flat-leaved parsley, are sold in many markets, especially where there is an Asian or Greek community. The seed (coriander) is sold both whole and ground, and is a major ingredient in curry powder.

History Coriander seed was mentioned in the Bible, where it was likened to manna, but its use goes back much farther in time. The herb was used both in cooking and medicine in the ancient European cultures, and in South America, India, and China many thousands of years ago. The Romans took it to Britain, where it was much used in Elizabethan times.

Characteristics The plant grows to a height of 24 inches, with a spread of 9 inches. The bright green leaves are fan-shaped and become more feathery toward the top of the plant. The flowers, which bloom from mid- to late summer, are small and white, formed in umbel-like clusters. The pale brown roots are fibrous and tapering, shaped rather like a carrot.

Growing tips A hardy annual, the plant is easy to grow from seed planted outdoors in late spring. It likes a light, well-drained soil and plenty of sun. Harvest the seed as soon as it starts to ripen, and hang the stems in paper bags to dry the seed.

How to use The leaves do not dry well, but may be frozen. They are used in curries or ground to a paste with olive oil as a covering for roast lamb in marinades. Cilantro has no cosmetic value but the essential oil is used extensively in the perfume industry and in powders, soaps, and toilet waters. The sweet and musky essential oil is expressed from the green berry whilst the dried is used for scenting dusting powder.

Medicinally, the leaves are rich in oils giving the herb strong digestive properties and the ability to stimulate the appetite. Cilantro has been used against bronchial congestion and has good antifungal and antibacterial actions.

How to take 15 drops of liquid tincture made from the leaves can be taken twice daily.

hawthorn *Crataegus laevigata*

A familiar sight in May with its white, scented flowers in the hedgerows, hawthorn has been used since the Middle Ages as a heart remedy. Widely distributed in Europe, hawthorn is a tough plant that makes a good windbreak, withstanding coastal conditions and urban pollution.

- **History** Both good and bad superstitions surround it. Bunches were tied outside the house to ward off witches and storms. A herald of summer, it was also an omen of death as in pagan times the "king and queen of the May" would be killed as human sacrifice. Also there is a whiff of decay in the scent. Another name was the "bread and cheese" hedge as the young leaves were a good addition to a ploughman's lunch. *Crataegus* comes from the Greek *kratos* that refers to the strength of the timber.
- **Characteristics** An extremely hardy deciduous shrub or small tree, densely branched with sharp thorns and lobed obovate leaves which are a shiny fresh green. Scented white, with red stamens that appear in late spring and early summer, followed by dark red, egg-shaped fruits in clusters. Height 15 to 20 feet. Spread 18 feet.
- **Growing tips** This is an easy tree— any soil, any aspect, though for fruit and flowers a sunny situation is better. Buy a small tree or grow from seed.
- **How to use** Medicinally, herbalists have used the berries from this plant for digestive problems for many years. It also has a tonic effect on the heart and circulation. The heart beat is strengthened, helpful in cases of a failing heart, while the blood vessels are dilated (opened up) which reduces the blood pressure and strain on the heart.

Hawthorn also has a diuretic action (increasing the flow of urine) on the body, ridding it of excessive fluid commonly collected by those with heart problems. The hawthorn berries are rich in vitamin C and bioflavinoids, essential factors for blood vessel strength and health.

- **How to take** Take 20 drops of liquid extract twice daily or try using capsules of the dried herb following manufacturer's instructions.

cone flower *Echinacea*

The cone flower makes a striking show from midsummer through to autumn when many other summer flowers have gone over. It has bold daisy-like flowers that are good for cutting for the house. The cone flower is native to central and northeastern North America growing in prairies and open woodland.

■ **History** The name comes from the Greek *echinos*, meaning hedgehog or sea urchin because the daisy flowers have prickly centers. It has a long history among the Native Americans. The tribes used it to soothe stings and for insect and snakebites. A piece of fresh root was chewed to alleviate toothache. It was also used to treat coughs and colds.

■ **Characteristics** *Echinacea* is a hardy vigorous upright perennial growing to 4 feet with large, daisy-like magenta or purple flower heads with a prickly orange central cone and lance-shaped leaves, sometimes toothed. It flowers from late summer through autumn. There are nine species. The three main ones are *Echinacea purpurea*, *E. pallida*, and *E. angustifolia*. There are some good cultivars including Robert Bloom, which has crimson flowers, and White Lustre, with snowy petals and a brown cone.

■ **Growing tips** Grow from seed or divide the roots in spring or autumn. *E. purpurea* can be sown outside. Plant in sun or dappled shade in light loamy soil.

■ **How to use** Medicinally, *Echinacea* is probably one of the most commonly used herbal extracts used today. In Germany the liquid extract is often referred to as "resistance drops," owing to the powerful immune stimulating effect of *Echinacea* during the winter months.

For all problems relating to bacterial, fungal, or viral infection *Echinacea* should be the first herb of choice.

■ **How to take** In acute illness, up to 40 drops of a liquid extract should be taken every 4 hours. Half the dose for children. For prevention take 10 to 15 drops daily. Alternatively try taking 1 to 2 tablets (100 mg) of dried extract daily.

As a cream for cuts and grazes, apply as required.

fennel *Foeniculum vulgare*

With its umbels of minute yellow flowers and dark green or bronze wispy leaves, fennel is a decorative addition to a herbaceous border, where, because of its size, it makes a good background plant. For centuries the herb has been associated in cooking with fish and used medicinally as a digestive. The seeds are chewed as a breath freshener.

■ **History** Fennel is native to southern Europe, and was extensively used by the Romans. Its use in England was widespread before the Norman Conquest. Its partnership with fish was so well established, on fast days poor people are said to have eaten the fennel without the fish.

■ **Characteristics** The plant dominates any border, growing to a height of 5 feet, with a spread of 30 inches. The stems are pale green, multi-branched, and ridged. The leaves, soft and frondy, are formed like giant hands and have an anise-like flavor. The seeds—which are flat, ridged, and oval—form in late summer and have a more pronounced taste.

■ **Growing tips** Seed may be sown "on site" in spring, and the developing plants should be thinned to 24 inches apart; or existing plants—hardy perennials—may be increased by root division. They like a well-drained soil and a sunny position.

■ **How to use** The leaves are used with pork, veal, and fish, in fish stock, sauces and stuffings, and in salad dressings. The dried stalks are placed under broiled or grilled fish to impart flavor. The seeds are used as a spice, particularly in bread, scones, and biscuits.

Use the chopped leaves to make soothing, stimulating infusions for masks, steamers, cleansers, and toners that are particularly kind and beneficial to older skins. An infusion of the seeds makes a gentle and restorative eye wash.

Medicinally, fennel aids in the digestive process and can be soothing for cases of colic and abdominal discomfort. The famous remedy for infantile colic, gripe water, is based on a fennel extract mixed with dill. Fennel also has a mild diuretic action (increases the flow of urine) and a cleansing action on the kidneys.

■ **How to take** Take 20 drops of a liquid tincture just after a meal.

■ **Warning** Avoid in pregnancy.

maidenhair tree *Ginkgo biloba*

*The **Ginkgo** is said to be the most ancient tree on earth and is sacred in China. It is a magnificent specimen tree often grown in the West to line avenues as nothing adversely affects it, not even pollution. The leaves and seeds are used in Chinese medicine for asthma and coughs. It is grown for its beautiful foliage, and its curiosity as the "fossil tree."*

■ **History** The maidenhair tree has been found in fossil records predating mammals. Seeds were sent back from China and Japan in the eighteenth century. The name comes from the Japanese *Ginkyo* or "silver apricot." Native to central China, it is extinct in the wild though preserved and grown in temple gardens.

■ **Characteristics** A hardy deciduous tree reaching up to 130 feet. Conical when young, its branches form a broad crown with age. It has lobed, fan-shaped leaves up to 5 inches across on long stalks. They are spectacular in autumn, turning golden yellow. Male flowers, which are short yellow catkins, and female flowers, which come in pairs of tiny cap shaped ovules, are borne on separate trees. The fruits—which smell of rancid butter and contain edible nuts—only appear when trees are near to each other and after a long summer.

■ **Growing tips** Buy a small tree or propagate from ripe seed. Grow in any fertile soil in sun.

■ **How to use**
Medicinally, *Ginkgo* is well known for its stimulating effect on the circulation to the brain and periphery but it has also been effectively used for treatment of asthma, allergic inflammatory conditions, Raynaud's disease, and varicose veins.

■ **How to take** Try using 20 drops of liquid extract twice a day. For a stronger response use a standardized extract in capsule form.

licorice *Glycyrrhiza glabra*

Licorice is a tall and graceful plant with green leaflets and pale blue or violet pea-type flowers. An important medicinal herb in ancient times and still used in many medical preparations, it is equally valuable for confectionery and flavoring. Native to the Mediterranean, it is grown commercially throughout the temperate zones.

■ **History** Licorice was mentioned on Assyrian tablets. Some was found in the funeral chamber of Tutankhamun. Roman legionnaires chewed it, as did Napoleon, we are told, to calm his nerves on the battlefield. It reached Europe in the fifteenth century. The Dominican friars at Pontefract Castle in Yorkshire, England made it popular with their famous Pontefract cakes and the licorice lozenges known as pomfrets. The name comes from the Greek *glycys* or sweet and *rhiza*, root.

■ **Characteristics** Licorice is a hardy perennial with stoloniferous roots, downy stems, and pinnate leaves, with nine to 17 leaflets. Pale blue to violet pea-flowers on loose spikes are followed by oblong pods up to 1¼ inches long. Height 4 feet. Spread 36 inches.

■ **Growing tips** Licorice likes a deep rich sandy soil in a sheltered sunny spot. Plant pieces of root, each with a bud or two, about 36 inches apart when the plant is dormant. Seed sown in autumn or spring is an alternative though this method can be slow. Once established, licorice can be difficult to eradicate.

■ **How to use** Medicinally, the root can help Addison's disease, asthma, and has a powerful detoxifying action on the liver. However, the sodium content of this herb makes it unsuitable for pregnant women or those with high blood pressure.

An estrogen-like action has been noted that makes it a good remedy for menopausal problems. Taken for stomach problems, licorice has a great healing power on the lining of the stomach. The cells are stimulated to multiply and ulcers can be cured within 1 to 2 months of regular use.

■ **How to take** Chew 2 to 3 tablets (100 mg) with each meal.

■ **Warning** Avoid in pregnancy, high blood pressure, kidney disease or in those taking the heart drug, digoxin.

St John's wort *Hypericum perforatum*

St. John's wort is an attractive plant for the garden with its starry yellow flowers that bloom all summer. The oil contained in the flowers and leaves smells of incense. It has been surrounded with myth and mystery since the earliest times. St John's wort is native to Europe and Asia in open woodland but now grows wild in Australia and America.

History The Greek name *hypericum* means "over an apparition"—possibly referring to its attributed power to dispel evil spirits. The English name comes from John the Baptist, the red pigment in the flowers (*hypericin*) signifying his blood. It is in full flower on June 24—both Midsummer's Day and his Saint's day. In the Middle Ages it was thought to drive away ghouls, witches, and storms. If gathered while the dew was on it, it would help you to find a husband or to conceive. It was used by the Crusaders to heal wounds.

Characteristics *Hypericum perforatum* is a perennial with stiff angular stems. The ovate linear mid-green leaves are covered with translucent dots containing the oil, which is also found in the flowers. They are star shaped and bright yellow, usually in clusters of three and last throughout summer. Height up to 36 inches.

Growing tips Grow in any well-drained soil in sun or part shade. The easiest way to propagate is by division, or sow seed in autumn or spring. (Harmful if eaten, poisonous to animals. Causes skin allergies in sunlight.)

How to use Medicinally, the antidepressant actions of this herb were found to be due to the high concentration of a substance called *hypericum*. Another interesting aspect is its ability to stop the multiplication of certain viruses (retroviruses) which may be of benefit to AIDS sufferers.

How to take Take 20 drops of liquid tincture three times a day or use a standardized capsule containing the powdered extract. When used in oils and creams it may cause problems when the skin is exposed to sunlight.

hyssop *Hyssopus officinalis*

This decorative and long-lasting herb is an attractive one to grow. Native to southern Europe, the Near East, and southern Russia, it is a garden escape in the United States. It has a slightly bitter flavor with overtones of mint, and was so widely known in ancient times Dioscorides wrote that it needed no description.

History Hyssop was used in all the Mediterranean countries in pre-Christian times, and is mentioned in the Bible. In his *Herball*, Gerard records that he grew all kinds in his garden; while Culpeper recommended it, boiled with figs, as an excellent gargle.

Characteristics The plant grows to 24 inches tall, with a spread of around half that. The leaves are about 1 inch long, pointed, oval and dark green. The flowers, which bloom from midsummer to mid-fall, are mauvish blue, ¼ inch long, and carried in long, narrow spikes. The stems, flowers, and leaves all give off a strong aroma.

Growing tips The plant, a sub-shrub, enjoys a dry, well-drained soil, and a sunny position. It can be grown from seed planted in spring, from root divisions, or from tip cuttings taken before flowering. The shrub needs protection in severe winters, and may need replacing every five years or so.

How to use The fresh or dried leaves and flowers may be added to soups, ragouts, casseroles, and sausages. Fresh leaves may be used sparingly in salads. The herb is an ingredient of Chartreuse liqueur.

Use the leaves and flowering tips to make healing infusions to treat acne. A healing oil may be obtained by pounding the flowers and leaves in enough sunflower or olive oil to cover, leave for one week in a warm place, then strain and use in a non-greasy moisturizer to treat spots and pimples. The highly aromatic essential oil is used in creams and body rubs to relieve aches and pains. Commercially hyssop is used to make eau de Cologne.

Medicinally, the herb has purification properties especially in cases of treating lung infections. Hyssop also has the ability to stabilize a low blood pressure and prevent the dizzy spells experienced by those sufferers as they rise from a sitting or lying position. An external application can be used for the treatment of minor cuts and bruises.

How to take Take 2 tablets (50 mg) of dried herb twice a day or try 15 to 20 drops of liquid extract taken twice a day.

orris root *Iris germanica*

Spectacular as the flowers are, it is the root or rhizome of the Florentine iris that is the valuable part of the plant. The name orris derives from the Greek word for rainbow, indicating the range of flower colors.

■ **History** Orris root originates from southern Europe, and became naturalized in Iran and northern India. It has been identified in a wall painting of an Egyptian temple dating from 1500 BC. It was at one time used as a purgative, but is not now used medicinally.

■ **Characteristics** The long and slender plant grows to a height of 36 inches, while its straight, fleshy, and erect stems are wrapped in long, pointed, sword-shaped leaves. The flowers are about 4 inches across and may be white, tinged with mauve or with a yellow beard. The bulbous and fleshy rhizome is white under the skin and smells strongly of violets. The plant has small fibrous roots.

■ **Growing tips** The rhizomes are divided in late spring and should be taken with a bud or shoot in place. They prefer deep, fertile, well-drained soil and a position in full sun. They should be planted half above and half below the soil, and divided every

four or five years. For dried orris root, lift the rhizomes in fall and hang them in a warm place. The fragrance develops as the rhizomes wither and dry.

■ **How to use** The deliciously violet-scented powder obtained from the ground root of *Iris* is used extensively in powders for the face and body and in potpourri, tooth powder, and bath salts. Oil of orris is extracted by distillation of the root under steam. It is the exact odor of the violet flower and is used in the most expensive perfumes.

Medicinally, orris is very rarely used but it was traditionally taken for bronchitis and asthma.

jasmine *Jasminum officinalis*

An ornamental garden-worthy climber, jasmine grows wild from the Caucasus to China. It is grown worldwide for its exquisite scent as is the royal or Spanish **Jasminum grandiflorum***. The Arabian* **Jasminum sambac** *is the variety used for flavoring jasmine tea.*

- **History** The name *Jasminum* comes from the Latin rendition of the Persian *yasmin*. It was known to the Assyrian kings and was recorded as growing in the Islamic desert gardens of the eleventh century. It came to Europe in the sixteenth century when it was taken up by the perfumery trade.

- **Characteristics** A vigorous, deciduous, twining semi-evergreen climber with green stems and elliptic leaves with three to nine leaflets. Fragrant white flowers in branched cymes throughout summer, followed by black fruits. Height up to 33 feet.

- **Growing tips** Grow in any fertile, well-drained soil. It is happy in sun but grows well on a north-facing wall. Thin as necessary after flowering. Propagate with greenwood cuttings in late summer.

- **How to use** Pure essential oil of jasmine is one of the most evocative of perfumes and one that truly lifts the spirits. Jasmine is a cornerstone of many perfumes in which it supplies the "middle note." Nowadays it is used almost exclusively in the perfume industry.

Medicinally, jasmine has been used for the treatment of sunstroke, fever, irritant dermatitis and infective illness. Emotional upsets and headaches respond well to a dose of jasmine. It is regarded as an aphrodisiac when applied to the body in its oil form.

- **How to take** Take a cup of jasmine tea 2 to 3 times a day. Mix 5 drops of the essential oil of jasmine with carrier oil such as almond oil and apply to the skin as needed.

bay *Laurus nobilis*

Bay leaves are among the most versatile of herbs, and the plants, if regularly clipped, rank among the most decorative of shrubs. The glossy and sweetly scented leaves are indispensable in both French and Mediterranean cooking, traditional ingredient in **bouquet garni**, *and a "must" in marinades,* **court bouillon**, *stocks, and relishes.*

■ **History** The bay tree came originally from Asia Minor, and was established around the ancient cultures of the Mediterranean. A crown of laurel leaves presented as a symbol of wisdom and victory in ancient Greece and Rome, is the origin of the circlet of leaves worn by victorious motor racing drivers today. The French word *baccalaureat*, for examinations, and the term "bachelor," for academic degrees, both derive from the Latin for laurel berry, *bacca laureus*.

■ **Characteristics** Bay leaves are flat, pointed, oval, about 3 inches long, dark green, and glossy. They retain a balsamic scent, and the wood, too, is strongly aromatic. The stems are tough and woody, and have a gray bark. The flowers, which appear in late spring, are small, yellow, and rather insignificant.

■ **Growing tips** Propagation is by heel cuttings, taken in early summer and kept under cover. Young trees are best planted out in spring; they like good, well-drained soil and a sunny, sheltered position. Harsh winters can kill them if they are left too exposed. For this reason, bay is often grown in pots or tubs.

The trees can be clipped into neat topiary shapes, the sphere being the most traditional.

■ **How to use** Bay leaves are used in all branches of cooking—in soups, stocks, syrups, sweet and savory sauces, and as a garnish.

Add them to the bath with a tablespoon of vinegar to relax muscles or include them in a bath bag. A small teaspoon of this oil to 9 tablespoons of alcohol and 1 tablespoon of mineral water, rubbed into the scalp before shampooing is a good conditioner.

Medicinally, the yellow-colored oil is responsible for its therapeutic actions. It has traditionally been used for the treatment of alopecia along with other problems such as flatulence, indigestion, and rheumatism.

■ **How to take** 20 drops of liquid tincture made from the leaves taken twice daily

lavender *Lavandula angustifolia*

Lavender is a traditional cottage-garden plant, its gray-green spiky foliage and spires of mauve-blue flowers providing color throughout the year. It is native to the Mediterranean and grows in profusion in the sun-baked Maquis region of southern France.

■ **History** The Greeks and Romans used this highly aromatic plant to make perfumes and ointments. Since the Middle Ages, the dried flowers have been one of the main ingredients of potpourri and fresh sprigs were included in herbal bunches, known as "tussie mussies," to mask unpleasant household odors and ward off fevers.

■ **Characteristics** The plant may grow to a height of 36 inches, but there are dwarf forms for edging. The stems are thick and woody, and become straggly if left unpruned. The leaves are long (about 3 inches), spiky, and very narrow. The tiny tubular flowers are carried on long spikes in thick clusters. The fibrous roots are shallow and wide spreading.

■ **Growing tips** The plants like a dry, well-drained, and preferably stony soil and a warm, sunny position. They should be lightly pruned in spring. Propagation of this evergreen shrub is by cuttings taken in spring or late summer.

■ **How to use** Fresh lavender flowers may be used to flavor syrup for jellies and fruit salad, and milk and cream for desserts. They may also be candied to decorate cakes.

Lavender is used extensively to make antiseptic, perfumed infusions, tinctures, herbal oils, cosmetic creams and lotions, toilet water, powders and deodorants, insect repellent and potpourri.

Medicinally, it is said to have antidepressant and mood-elevating effects, and has been used internally for the treatment of digestive problems, anxiety, rheumatism, irritability, and tension and migraine headaches. Applied externally, lavender is of benefit to burns and rheumatic pains.

■ **How to take** As an external application, use 5 drops of the essential oil mixed with enough base oil (almond) to treat the affected area.

lemon balm *Melissa officinalis*

Balm is an attractive herb with yellow or variegated leaves smelling strongly of lemons. It is a great addition to any garden since it maintains a strong attraction for bees. Indeed, it used to be said that a swarm of bees would never desert a hive if a lemon balm plant was close by.

■ **History** Lemon balm is native to southern Europe and has been cultivated for over 2,000 years. The Romans took it to Britain, where it was widely grown in the Middle Ages, during which time Melissa honey was popular as a sugar substitute.

■ **Characteristics** The plant is a vigorous grower that will readily spread through the border. It reaches a height of 36 inches, with a spread of 24 inches. The oval, almost heart-shaped leaves have slightly serrated edges and a pronounced network of veins. The flowers, which bloom from mid- to late summer are small, white, and insignificant.

■ **Growing tips** Grow lemon balm in sunken pots or other containers to prevent unwanted spreading. The plant is a perennial that dies down in winter. It is grown from seed sown outdoors in spring, or from root division. It prefers a moist, fertile soil and partial shade.

■ **How to use** The fresh leaves may be used for salads, candied for cake decoration, and used to garnish fish and other dishes. Add them to summer drinks and fruit salads.

The leaves can be used to make soothing astringent infusions for cleansers and can be added to a bath bag for scent and effect. The dried leaves add a pleasant lemony fragrance to potpourri. Essential oil of *Melissa* can be added to bath and massage oils. A few drops in a base oil is an effective insect repellent.

Medicinally, this lemon-scented herb has powerful antiviral and antibacterial effects that can be helpful in the treatment of recurrent cold sores. An application of a good *Melissa*-based cream to the area just as the cold sore is felt can often prevent it erupting. Taken internally, the herb helps with nervous problems and depression. Those who suffer from

panic attacks and heart palpitations may find the extract helpful.

■ **How to take** As a cream, apply three times a day. The oil can be mixed with almond oil and massaged in. Take 15 drops of a liquid extract twice daily.

mint *Mentha*

There are many species and types of this most popular of family, the Labiatae. Spearmint, or garden mint, is the most commonly grown. **Mentha x piperita***, peppermint, is the main medicinal herb of this genus. Most species are native to the Mediterranean region and western Asia, and now grow wild throughout North America.*

- **History** Mint was used extensively by the Greeks and Romans. And it was the Romans who introduced both spearmint and mint sauce to Britain.
- **Characteristics** The plants may grow to a height of over 18 inches. They have tough, vigorous roots and stems, which creep beneath the ground and establish new plants along the way. The small, bluish-mauve flowers, which tend to bloom late in the summer, are borne in clusters on cylindrical spires.
- **Growing tips** The foliage dies in autumn, which is the time to trim the plants and surround them with a mulch. If the plants are attacked by rust they should be burned and fresh ones started in another area.
- **How to use** The main constituent in peppermint is the compound menthol which gives it its cool and tingly taste and feel and which makes it invaluable for toothpastes, after shaves, soaps, and

bath essence. However, the essential oil of peppermint should be used with caution and never be used undiluted at any time. Peppermint oil rather than essential oil of peppermint is a culinary preparation which can be used in some cosmetics, e.g. lip balm.

Medicinally, peppermint has a history as a decongestant and antiseptic agent. The plant's oils have an antispasmodic action on the muscle of the gut making it useful for irritable bowel syndrome.

As a remedy for nausea and morning sickness in pregnancy, an internal dose of peppermint is very safe. The external application of the oil helps in controlling muscular discomfort.

- **How to take** 2 to 3 capsules (2 ml of oil per capsule) between meals to relieve bowel spasms. As a tea, take 1 cup twice daily.

bergamot *Monarda didyma*

*The bergamots are native to North America. The plant has a pleasant smell of oranges and is strongly attractive to bees. Besides the most common red bergamot, there is wild bergamot (**M. fistulosa**), native to southern Canada and the northern United States, and lemon bergamot (**M. citriodora**), which, too, has a strong citrus aroma.*

History The Oswega tribe used bergamot, and Oswega tea was made by colonists at the time of the Boston Tea Party in order to boycott British imports.

Characteristics The plant, a herbaceous perennial, will grow to a height of 36 inches, with a spread of over 12 inches. The fibrous roots form a thick, dense block. The dark green leaves, which may be tinged with red, are hairy and up to 16 inches long. The flowers, up to 2 inches long, are borne in thick clusters at the top of the stem from mid- to late summer.

Growing tips The plants may be grown from seed planted in spring or from root divisions taken in spring or fall. They like a sunny position and moist but well-drained soil.

How to use Fresh leaves may be used sparingly in salads, fruit salad, and fruit drinks, and fresh or dried leaves can

be made into a refreshing and relaxing tea that is said to be soporific. The dried leaves lend a pleasantly citrus aroma to potpourri.

Bee balm or red bergamot has no cosmetic values but should not be confused with bergamot oil which is a cosmetic oil extracted from the rind of a bitter orange tree and which is used in toilet waters and body oils. At some time the confusion may have arisen because red bergamot has a distinct orange smell.

Medicinally, the leaves can be used to help relieve nausea, flatulence, and menstrual pain. Bergamot can also be used in cases of catarrh as a steam inhalation.

How to take Take 15 to 20 drops of a tincture twice daily, alternatively drink bergamot tea 2 to 3 times daily.

sweet cicely *Myrrhis odorata*

Sweet cicely grows wild in northern Europe, and provides good visual value in a border or herb garden. With its large, bright green, lacy leaves and mass of creamy-white flowers, it makes a perfect back-of-the-border plant. The whole plant of this herbaceous perennial is fragrant, with a mildly aniseed aroma.

- **History** In his *Herball*, Gerard stated that when the seeds were eaten with oil, vinegar, and pepper they "exceed all other salads." The botanical name of the plant is thought to recognize that the leaves have an aroma similar to that of myrrh, and the popular name indicates its sweet flavor.

- **Characteristics** The plant is so decorative, neat, and tidy, it easily earns a place in a flower border, where it may spread to 36 inches across and grows to twice that height. The thick, hollow stems are deeply ridged, while the leaves, pale on the undersides, are toothed and fern-like, and up to 12 inches long. The flowers appear in late spring and early summer, are attractive to bees, and number among the prettiest of the umbellifers. The roots are long, thick, and fleshy, and white inside a light brown skin. The seeds, which may be up to 1 inch long, are brownish black, long, narrow, and sharply pointed.

- **Growing tips** The seeds are slow to germinate, and so it is best to grow the plant from root division; a small piece of root with an eye is all you need. Plant it 2½ inches deep in spring where it is to grow. Transplanting is difficult, because the roots are deeply penetrative. It enjoys a deep, moist soil and partial shade.

- **How to use** The leaves may be used fresh in salads and fruit salads, and chopped into other fruit dishes such as pies and compotes. The peeled roots can be boiled and eaten as a vegetable, accompanied by a white sauce or vinaigrette dressing. The seeds are used in the making of the well-known Chartreuse liqueur.

  Medicinally, this herb is very rarely used. Traditional remedies for diabetes may contain sweet cicely.

catmint *Nepeta cataria*

Catmint, the herb so attractive to domestic cats, has few culinary uses. With its gently curved spikes of heart-shaped, gray-green leaves and clusters of white or pale blue flowers, it is, however, an attractive addition to a border.

- **History** The plant, an herbaceous perennial, is native to Asia and Europe, and was widely used in self-help medicines. In his *Herball*, Gerard recommends it for colds, coughs, chest complaints, and nervousness.

- **Characteristics** The plant grows to a height of up to 3 feet, with a spread of 15 inches. It has a straggly habit, and is liable to be squashed flat by cats rolling on it. Indeed, to preserve plants from this fate, it may be necessary to protect them with wire netting.

- **Growing tips** Catmint can be grown from seed planted in spring or summer in good fertile soil in partial shade, or by root division or cuttings taken in spring.

- **How to use** The fresh leaves, which have a very strong aroma, can be used sparingly in salads.

  The pungent smell of catmint inhibits its use in cosmetic preparations, however a handful of the leaves in a bath will relieve an itchy skin while an infusion, used as a hair rinse, soothes scalp conditions and promotes growth. Two tablespoons each of dried rosemary and catmint simmered for 10 minutes in 2 cups of water, steeped for two hours and then strained and added to ½ cup of cider vinegar makes a good conditioning final rinse for dark hair.

Medicinally, catmint tea can act as an effective tonic for the stomach in cases of infantile colic and diarrhoea. External applications are helpful for minor cuts, abrasions, and bruising.

- **How to take** Adults: 20 drops of liquid tincture twice daily.

  Babies and children: 5 to 10 drops added to warm water. Give this between meals.

evening primrose *Oenothera biennis*

An American native, the evening primrose is either regarded as a weed or treasured as a border plant for its phosphorescent, fragrant flowers that open at dusk. The American Flambeau Ojibwe tribes used the plant to treat bruises, skin complaints, and asthma. Modern research is discovering other important medicinal properties.

- **History** Seeds of evening primrose were brought from America to Padua Botanic Garden in Italy in 1619. Theophrastus, the ancient Greek physician, is thought to have given it the name *Oinos*, wine and *thera*, hunt. Possibly it was used as a hangover cure.

- **Characteristics** An erect biennial reaching 3 to 5 feet. The first year there will be a rosette of leaves and in the second year it will shoot up and flower. The mid-green leaves are lance-shaped, slightly toothed, and sticky, and grow in rosettes. The flowers are bowl-shaped and fragrant, about 2 inches across, ageing from pale yellow to gold. They open in the evening to attract moths and keep coming from midsummer to autumn.

- **Growing tips** Sow seed in late spring in well-drained soil and a sunny position. Thin to 12 inches. It will self-seed freely once established.

- **How to use** Oil from capsules of evening primrose oil can be added to night and day creams, especially those formulated for stressed or ageing skins. Two capsules per pot are usually enough. It can also be applied direct as an emergency measure to blemishes.

Medicinally, the oil extracted from this plant can stimulate the healthy functioning of all body cells in the balanced output of hormones. This makes the oil useful in cases of premenstrual syndrome and fibrocystic breast disease. It is interesting to note that schizophrenia has responded well to evening primrose supplements, the mechanism behind this is unknown.

- **How to take** For menopausal symptoms, use 2 to 3 capsules (500 mg capsules) every evening, with water only.

  For premenstrual symptom, take 3 capsules (500 mg capsules) every evening for about 14 days before the onset of the period.

  For children, use 250 mg of oil mixed in food daily.

  For cradle cap, massage enough to make the area supple.

  - **Warning** Avoid use in epileptics and migraine sufferers.

sweet basil *Ocimum basilicum*

With a pot of basil on the windowsill and a tomato plant in a window box outside, you have the perfect partnership for many summer salads and sauces, for basil and tomatoes go together in any combination. The herb is a half-hardy annual emanating from warm climates and is therefore a sun-lover.

■ **History** Basil traveled overland to Europe via the Middle East from India, where it was considered sacred by the Hindus. A Belgian old wives' tale of the sixteenth century told that basil leaves crushed between two bricks would turn into scorpions, while Boccaccio has his heroine Lisabetta burying her lover's head in a pot of basil and watering it with her tears.

■ **Characteristics** According to type, a basil plant can have either a multitude of minute leaves or ones up to 4 inches long and almost half as wide. Glossy, smooth, silky, and highly aromatic, the leaves smell similar to cloves. The stems tend to be woody and straggly, and the flowers, in long spikes, are small, white, and tubular. They appear from midsummer through to the fall. The plants can reach a height of 24 inches.

■ **Growing tips** Basil is the ideal windowsill herb, but will grow outdoors if sheltered from the wind.

Sow seed shallowly in gentle heat in late spring and transplant toward midsummer, taking care not to disturb the roots. It likes a moist but well-drained soil and plenty of sun. Pinch out the tips of the shoots to prevent flowering and encourage bushy growth. Harvest by taking the larger, lower leaves first.

■ **How to use** In Italy and France, basil is used to make *pesto* or *pistou* sauce, in which it is crushed with garlic and pine kernels. The sauce may be served with spaghetti or stirred into soup.

The fresh and dried herb is antiseptic, sweet smelling, and relaxing and is used in bath and after-bath preparations. Frequently used in potpourri, it not only smells nice but also repels flies. The essential oil is very aromatic and is mainly used in soaps and perfumery.

Medicinally, basil can be taken for chills, colds, and influenza where it has a stimulant action. For the digestion, basil is of great help in cases of stomach inflammation and abdominal cramps associated with menstruation. Basil is sometimes referred to as St Joseph wort, not to be confused with St John's wort.

■ **How to take** Take 15 drops of a liquid tincture daily.

oregano *Origanum vulgare*

Oregano is a very close relative of marjoram, so much so there is some confusion in the cross-pollination of their names. This is the pungently aromatic herb of southern Italy, the one that is used, mainly in its dried form, to flavor pizzas and tomato sauces. Indeed, Greek cooks are convinced that oregano—rigani—is best used dried.

- **History** The plant originates from the Mediterranean region, where its pungency is in direct proportion with its quota of sun. It is a traditional ingredient of Mexican chili powder, and has long been used as a flavoring for chili sauces and chili beans.
- **Characteristics** This hardy annual grows to a height of about 8 inches, with woody stems and dark green leaves around ¾ inch long. The flowers, borne on long spikes, are small and white in color.
- **Growing tips** The plant demands a well-drained soil in full sun, though a poor, stony soil will be adequate. Plant seed in warm soil in late spring, or in mid-spring, in pots or seed trays under glass. Oregano does especially well in indoor mini-propagators placed on the windowsill.
- **How to use** The fresh leaves, which are sold in bundles in Italian and Greek markets, are useful additions to salads, soups, sauces, pâtés, and poultry dishes. Dried oregano is especially good with tomatoes, beans, eggplant, zucchini, and rice, and in dishes such as pilaf and risotto.

The leaves and flowers may be used in much the same way as sweet marjoram but are far less aromatic. The essential oil is hotter and spicy but it is very antiseptic and can be used to advantage in soaps and medicated bath oils.

Medicinally, wild marjoram has been used to treat a variety of health problems such as coughs and chest infections as well as colic and indigestion.

- **How to take** Apply the essential oil, mixed with almond oil as a carrier, to the chest region when suffering from coughs and colds.

Take 10 drops of a liquid extract twice daily for colic and indigestion.

sweet marjoram *Origanum majorana*

Sweet or knotted marjoram is highly perfumed and has thick trusses of dainty, white, pale mauve, or purple flowers, which make it one of the most decorative plants in the herb garden. It also represents good value since the leaves dry or freeze well for culinary use, and the flowers may be dried for long-lasting arrangements or potpourri.

■ **History** Sweet marjoram has been cultivated since ancient times. It is a native of central Europe, where it was grown for its many medicinal uses.

■ **Characteristics** The plant grows to about 10 inches high, with a spread of 8 inches. The stems are tough, woody, and inclined to be straggly, while the dark gray-green leaves are oval and up to ¾ inch long. The flowers are minute but plentiful, and are borne in clusters around the stem. They are produced from green pea-like buds known as "knots," which give the plant its alternative name.

■ **Growing tips** The plant prefers a moist, fertile soil in a sunny position that is sheltered from wind. Sow seed in late spring when there is no risk of frost. To aid germination, which in cold soil can take a month, it is a good idea to warm the soil first by protecting it with cloches.

■ **How to use** Add fresh leaves to casseroles just before serving to

retain the full flavor. They can also be used in sauces, stuffings, sparingly in salads, in egg and cheese dishes, and in fruit salads.

Sleep is one of the greatest aids to good looks and a herb pillow well stuffed with marjoram will ensure a good night's rest. An infusion of the leaves and flowers or a few drops of essential oil added to rosewater is a delightful skin toner. The exceptionally aromatic essential oil can also be used in bath and massage oils to combat tiredness.

A few drops massaged into the scalp gives dry hair a gloss and prevents hair loss.

ginseng *Panax ginseng*

Ginseng is a humble looking woodland plant credited with astonishing powers. It has been used in Chinese medicine for 5,000 years. Native to the dense mountain forests of Asia, ginseng is now rarely found in the wild. It is cultivated commercially in Korea, Russia, Britain, and in the USA—mostly in Wisconsin.

■ **History** The name comes from the Chinese *Jin-chen* or "manlike" as the roots are rather like a crude doll man with arms, body, and legs. *Panax* is from the Greek *panakos* or panacea. It is an ancient Taoist tonic herb that is used as a "vital essence" (*qi*) in Chinese medicine. Though introduced to Europe in the ninth century it was not taken up there until the 1950s.

■ **Characteristics** A hardy perennial with upright stems bearing whorls of five divided leaflets. Red berries follow umbels of green-yellow flowers in spring and summer. Height 28 to 36 inches. Other medicinal species of ginseng include *Panax japonicus*, found wild in central areas of Japan, *Eleutherococcus senticosus* from Siberia, and *P. quinquefolius,* the American ginseng which grows in Canada and the eastern United States. *P. pseudo-ginseng* was used by the Vietcong in the Vietnam War to speed healing of gunshot wounds.

■ **Growing tips** Germination is erratic. Sow seed in woodland-type soil—well-drained sandy loam with leaf mold.

■ **How to use** When taken internally, ginseng acts as a general tonic by stimulating the central nervous system. For the treatment of stress and chronic fatigue the stimulant effect may give relief but should not be taken daily.

Ginseng has been shown to reduce the blood concentrations of both glucose and cholesterol as well as stimulating resistance to disease.

The active agents in Siberian ginseng *E. senticosus* are similar to the *Panax* form but are considered to be less potent. Siberian ginseng can be taken for longer periods than *Panax* and is thought to be better suited to an extended period of treatment for stress.

■ **How to take** Use 2 teaspoons of ginseng elixir daily or capsules containing a commercially prepared dried extract.

sweet-leaved geranium *Pelargonium*

Pelargonium species originated in South Africa. Different varieties have different aromas: there is the lemon-scented P. crispum minor: apple-scented P. odoratissimum; oak-leaf-scented P. quercifolium; rose-scented P. graveolens and P. radens; nutmeg-scented P. fragrans; peppermint-scented P. tomentosum, and many others. The flowers have no smell.

■ **History** Tradescant, the gardener of Charles I of England, "discovered" the plants, and grew a number of varieties in the royal greenhouses. One of the first to be brought to England was *P. triste*, which numbers among the few species that have scented flowers as well as foliage.

■ **Characteristics** The plants have dark green, pale green, or green-and-cream variegated leaves, which may be deeply cut or frilled and may vary in size from ½ inch to 3 inches across. The five-petaled flowers are borne in clusters and are long-lived. Height varies considerably, and may be between 12 and 36 inches. The stems are tough and woody.

■ **Growing tips** Pelargoniums are grown from tip cuttings taken under cover in spring and summer. They like a good well-drained soil, plenty of sun, and protection from cold. Grown indoors, they require plant food once a week to encourage full leaf growth. The plants should be cut back in winter to avoid becoming straggly.

■ **How to use** The fresh leaves may be infused in milk, cream, and syrups for desserts, sorbets, and ices; chopped into softened butter for sandwiches and cake fillings, and used extensively for garnishing.

Use the flowers and leaves or the essential oils to add to healing and rejuvenating creams for oily and ageing skins. The essential oil is strongly floral and is used not only in skin preparations but in bath oils, massage creams, soaps, toilet waters, and perfumes and also as an insect repellent.

Medicinally, the dried leaf is not used in Europe but in Africa certain species of *Pelargonium* are used to treat diarrhoea.

parsley *Petroselinum crispum*

Parsley, with its deep green, frilled or curly leaves, is one of the best-known and most widely used herbs, as much for garnishing as for cooking. Neapolitan parsley, whose flat leaves are reminiscent of cilantro, is less decorative, has a sharper flavor, and is easier to grow.

■ **History** A native of the eastern Mediterranean region, parsley was first recorded in a Greek herbal as long ago as the third century BC. It was used in ancient Rome in cooking and for ceremonial purposes.

■ **Characteristics** The plant grows to a height of up to 18 inches, with a spread of 10 inches. The stems, which are also aromatic, are green and supple, the leaves curled or flat. The flowers, which appear in the plant's second year, are yellow-green.

■ **Growing tips** Outdoors, seed is best sown in late spring, summer, and early fall in warm soil where it is to grow. Soaking the seed in lukewarm water speeds germination, while pouring boiling water along the seed drill both warms the soil and provides the necessary moisture. In winter, the plants should be protected by cloches or dug up and potted for overwintering indoors.

■ **How to use** Parsley has its culinary uses in nearly every category of food, in preparing soups, sauces, in marinades, and with meat, poultry, fish, and vegetables. Often a sprig of parsley is all that is needed to present a dish attractively.

Use the crushed leaves or their extracted juice for healing poultices or lotions, to make infusions for cleansers and creams to treat oily skins or skins with thread veins. Has anti-dandruff and deodorizing properties.

Medicinally, parsley is rich in vitamins A and C and contains substances that help to reduce allergic reactions. An internal dose can help with menstrual cramps, inflammation of the bladder and prostate, while in the stomach it can reduce colic and indigestion. The stimulant effect on the uterus makes this herb one to avoid during pregnancy, but once the baby is born it may help stimulate lactation and milk flow.

■ **How to take** Take 20 drops of liquid extract twice a day.

■ **Warning** Avoid in pregnancy.

kava kava *Piper methysticum*

Kava kava is a member of the pepper family but, unlike the other culinary peppers, it is grown for its roots rather than its berries. It has been used in parts of the Pacific at traditional social gatherings as a relaxant and in cultural and religious ceremonies. It is gaining popularity for its relaxing effects and as a pain reliever.

■ **History** Kava kava is native to the Sandwich and Fiji Islands and is part of the social life of the Polynesian people. They make an intoxicating liquor from the rhizome root, which is known as *Kava, Ava,* and *Kawa*. The traditional method of making the beverage is to chew the root and strain the juice through coconut husks. Small doses are taken to stimulate the appetite, relieve pain, and relax spasms.

■ **Characteristics** *Piper methysticum* is a tender tropical shrub growing up to 16 feet by 12 feet. It has reedlike, knotty branches. The broadly heart-shaped leaves have prominent veins and can be up to 8 inches across. The flowers are small and unisexual. The root can weigh several pounds.

■ **Growing tips** Peppers need a minimum temperature of 60°F. They like well-drained soil in shade and plenty of water. In hot countries they grow luxuriantly but in cooler ones they are grown much like tomatoes in a greenhouse. Propagation is by runners or cuttings. (There are legal restrictions in some countries.)

■ **How to use** Medicinally, kava kava is used to stimulate the nervous and circulatory systems. It is a good cure for insomnia and nervousness by enhancing restfulness. Kava kava also has the ability to reduce the pain associated with muscle spasm and arthritis, enhance mental awareness, and combat fatigue.

■ **How to take** Two tablets (100 mg) of dried herb daily or use 15 to 20 drops of a tincture.

rose *Rosa*

Rose water is an age-old ingredient in cooking, especially in Middle Eastern sweets. The rose hips, which have a higher vitamin C content than citrus fruit, are used for wines, cordials, and syrups. The essential oil "attar of roses" is a classic for cosmetics. The rose has been valued throughout history for medicinal use.

■ **History** The Apothecary's rose, *Rosa gallica* var. *officinalis*, is probably the oldest garden rose in history. The Crusaders brought it from Damascus to Gaul in the thirteenth century. The name "rose" is thought to come from the Greek *roden* or red after this rose, which, according to myth, sprung from the blood of Adonis. *R. x damascena*, came from Persia in the fourtheenth century by the same route. Chaucer knew the sweet briar or eglantine, *R. eglanteria*. The humble dog rose, *R. canina*, the wild rose of English hedgerows, has the best hips for culinary use.

■ **Characteristics**
Roses are hardy deciduous shrubs.
R. gallica var. *officinalis* has a neat bushy habit growing up to 36 inches, leathery leaves, and highly fragrant, semi-double pink to red flowers. *Rosa x damascena* has arching prickly stems, grows up to 7 feet and produces clusters of semi-double fragrant pink flowers. *Rosa eglanteria* is a vigorous and prickly rose with apple-scented leaves and single cup-shaped rose pink flowers, growing to 8 feet. *Rosa canina* is a wild and arching climber up to 10 feet with scented single flowers.

■ **Growing tips** Except for *Rosa eglanteria* which thrives in dry chalky soil, plant in fertile, neutral to acid, moist soil in sun. If you buy a bare-rooted plant in winter, prune back almost to the ground. After that prune out dead, weak or damaged wood in autumn. Feed in summer.

■ **How to use** Although this is the rose from which traditionally pure rose oil is extracted, the petals from strongly scented garden roses can be used to make rosewater infusions for a range of skin preparations. The powerfully floral essential oil is used in healing and rejuvenating cosmetics, bath and massage oils, and shampoos. Geranoil, the chief constituent of attar of Roses, is also present in lavender, pelargoniums, lemongrass and neroli.

rosemary *Rosmarinus officinalis*

An evergreen shrub, rosemary is best used fresh because once dried it loses much of its flavor. It is a pretty herb, with trusses of pale or bright blue flowers lasting, in the right climate, right through spring and summer. It is native to the countries bordering the Mediterranean, where it grows in profusion.

■ **History** The name means dew (*ros*) of the sea (*marinus*)—the sea being the Mediterranean. In the Middle Ages in Britain, sprigs of rosemary were dipped in gold and tied with ribbon as a keepsake for wedding guests, and the herb was a traditional gift on New Year's Day.

■ **Characteristics** In the proper conditions, the bush can grow to a height of 6½ feet and spread 5 feet across. The leaves are dark green on top and silver-gray on the underside, about 1 inch long and ½ inch wide. The two-lipped tubular flowers are small, about ½ inch long, and carried on long thick spikes. The stems become tough and woody.

■ **Growing tips** The plant is grown from tip cuttings taken in summer and set in sandy compost to take root. It prefers a light, well-drained soil and a sunny, sheltered position, and can be fan-trained against a wall, where it makes a most decorative spreading shrub. Pinch

out the tip of the main shoot to encourage bushy side growth, and protect from frost and snow in winter.

■ **How to use** Rosemary and lamb go together in many ways. Use rosemary in *bouquet garni,* sparingly with fish, and in rice dishes.

Rosemary is one of the great cosmetic herbs used traditionally in toilet waters. The leaves and flowers are used to make therapeutic and antiseptic infusions for all cosmetic preparations and the leaves can be used to make herbal oils to massage into the hair and body. The essential oil is very true to the scent of rosemary, but as it is sometimes adulterated, be sure to buy the best.

Medicinally, the oil of rosemary has strong antiseptic actions as well as being a powerful anti-inflammatory. Rosemary has been used internally for the treatment of depression, lassitude, migraine and tension headaches, poor circulation, and digestive disorders. For rheumatism and muscular aches and pains its external application can give symptomatic relief.

■ **How to take** Apply the essential oil, mixed with almond oil as a carrier, to the joints twice daily.

Take 10 drops of a liquid extract daily.

sage *Salvia officinalis*

Sage is an evergreen sub-shrub, though its leaves are not necessarily green. Some varieties have gray or gray-green downy leaves, and one has deep purple leaves and exceptionally pretty mauve-blue flowers. The flavor, which has faint overtones of camphor, is very strong in some types. The plant is a native of the Mediterranean region.

■ **History** Sage, which takes its name from the Latin *salvere*, "to save," has a long history as a medicinal plant . It was much in evidence in Roman times for medicinalpurposes, but was established mainly as a culinary herb in medieval England.

■ **Characteristics** The plant can reach a height of 24 inches, with a spread of 18 inches. The stems are woody, and the leaves, elongated ovals, can be 2½ inches long and ¾ inch across. The flowers, which appear in midsummer, are about 1 inch long, borne on long, curving clusters.

■ **Growing tips** Sage likes a well-drained soil and a sunny position. It can be grown from seed planted in spring, or from tip cuttings taken under cover in summer. The plant tends to become woody and straggly, so should be renewed (by taking cuttings) every few years. Cutting off the stems before flowers form encourages leaf growth.

■ **How to use** Sage is a popular culinary herb and has an affinity with pork and cheese. It is also traditionally used with onion in stuffings.

The leaves and flowers are used to make astringent infusions for cosmetic use in cleansers and tonics to treat oily and problem skins. It also conditions and colors dark hair.

Medicinally, sage provides us with a readily available antiseptic agent, and its juice has anti-inflammatory and antiseptic properties. Smooth muscle (found in the internal organs) can be effectively relaxed by sage extracts. Sage can be taken to control excessive lactation. Menopausal problems may respond to the estrogen-like stimulation of sage and it has been used to assist fertility and aid digestive problems.

■ **How to take** Take 20 drops of the liquid extract twice a day.

■ **Warning** Avoid in pregnancy.

clary *Salvia sclarea*

A native of southern Europe, clary was introduced into Britain in the sixteenth century, when it was used in brewing and combined with elderflowers to give wine the flavor of muscatel. It is a close relation of sage and a decorative biennial that is usually treated as an annual.

History A sixteenth-century botanist said of clary that "it restores the natural heat, and comforts the vital spirits, and helps the memory." At the same time, the leaves were used in omelets, boiled with cream, and made into fritters to serve with orange or lemon juice.

Characteristics Clary grows to a height of 36 inches, with a spread of about 12 inches. It has oval, downy, dark green leaves, which can be up to 8 inches long. The flower bracts are about 2 inches long, and are carried on straight, branching spikes at the top of the plant; the bracts may be mauve, purple, white, or pink. The fragrance is somewhat balsamic, the taste bitter.

Growing tips The plant is grown from seed planted in spring, and will flower the following year. It requires a light soil and a warm, sheltered position if it is to survive the rigors of winter.

How to use The strongly aromatic leaves can be added sparingly in the preparation of soups, homemade wines and beer, or made into fritters.

Clary sage is used as a soothing complexion lotion. Drop a handful of freshly picked tops into 1 cup of boiling water. Cool, strain and use.

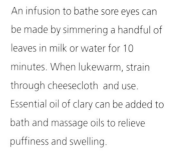

An infusion to bathe sore eyes can be made by simmering a handful of leaves in milk or water for 10 minutes. When lukewarm, strain through cheesecloth and use. Essential oil of clary can be added to bath and massage oils to relieve puffiness and swelling.

Medicinally, the leaves are used to make an infusion (a tea) and taken as a gargle for throat infections or applied to cuts and skin abrasions. It has been used in traditional herbal remedies to help ease vomiting and stimulate the appetite.

How to take Take 15 to 20 drops of liquid tincture twice daily or drink 2 to 3 cups of clary tea daily.

elder *Sambucus nigra*

In springtime, the aroma of the umbrella-like trusses of creamy-white elderflowers scents the hedgerows with a sweet aroma one longs to capture. The plant is common throughout Europe, western Asia, and North America, where a related species, Sambucus canadensis, *American elder, was used as a folk medicine by Native Americans.*

History Elder has attracted a strong folk history. It was thought that a tree planted outside a house kept witches at bay and protected the house from lightning. Cutting it back was said to bring bad luck.

Characteristics The tree can grow to a height of about 30 feet or more, with a spread of 9 feet, but many trees are much smaller. The leaves are dull dark green, about 4 inches long and finely toothed. The flowers are minute, highly fragrant, and carried in umbel-like clusters. The purplish-black berries are small and round, and hang in heavy trusses.

Growing tips The trees like a moist soil and plenty of sun if the flowers are to develop their maximum fragrance. Elders are grown from hardwood cuttings taken outdoors in the autumn.

How to use Elderflowers are traditionally used to flavor fruit and have a particular affinity with gooseberries. The berries—usually blended with apples—can be made into jelly and other preserves, while both flowers and berries create excellent wines. The flowers, blended with lemon and sugar, are used to flavor summer drinks and cordials.

In the past every part of the elder tree was used in potions and lotions for the body and hair and they are as effective

today as then. Use the flowers to make mildly bleaching infusions for softening cleansers, tonics, and conditioners for hands, face, and body. Use the leaves to make healing decoctions for blemished, sun or wind burned skin. Use the flowers to lighten and condition blonde and graying hair and the berries to improve the color of graying dark hair.

Medicinally, elder can help speed a cold away by increasing the body temperature. When a cold is on its way taking a hot tea made from elder will induce sweating and this boosts the body's viral-killing ability. An external application can be of great relief to irritated skin.

How to take Make an infusion (tea) and drink 2 to 3 times a day. Alternatively take 15 to 20 drops of a tincture made from the leaves, bark, and flowers twice a day. Apply creams as required.

Warning The seeds from the elder can be toxic and should be avoided.

saw palmetto *Serenoa repens syn. S. serrulata*

Native to the scrublands of North Carolina to Florida Keys, Mississippi, and Louisiana, the berries of this palm-like plant have a long-held reputation as a sexual stimulant and aphrodisiac for both men and women. Saw palmetto is widely used for its therapeutic effects on urinary tract problems and for improving body strength.

■ **History** The Latin name comes from the American taxonomist Sereno Watson (1826-1892). The fruits were an important food for Native Americans and later the American settlers. They were considered to be a good general tonic.

■ **Characteristics** The saw palmetto grows between 36 inches and 13 feet and makes substantial thickets in the wild with its creeping underground roots. Long stiff leaves of about 36 inches grow in fan shapes. They are yellow or grayish-green, narrow and pointed, and sometimes waxy. Fragrant ivory-colored flowers grow in clusters. The fruits are just over an inch long, turning black when ripe. The fruit is brown inside with a pip.

■ **Growing tips** Though saw palmetto does not transplant well from the wild and is rarely cultivated in gardens, it can be grown as a house plant. It needs a minimum temperature of 50°F and it likes humidity and damp (but not soggy) soil. Propagate from seed and feed container plants every two weeks throughout summer.

■ **How to use** Medicinally, the plant's berries were traditionally used for their sedative and tonic properties but it was soon noted that when taken internally, the berries act as an effective remedy for male impotence as well as the age-related onset of prostate enlargement and associated cystitis. The berries appear to keep the swollen prostate stable and even help reduce the size in some cases.

It has poorly understood effects on the body's hormone system with its aphrodisiac properties being unsubstantiated.

■ **How to take** Take 20 drops of liquid tincture twice a day or use a standardized capsule containing the powdered extract.

milk thistle *Silybum marianum*

The milk thistle makes a dramatic spiky plant for the border or gravel garden with a tracery of white veins, followed by purple thistle flowers. The plant originates from the mountains and stony thickets of Africa, the Mediterranean, and central Europe. It has naturalized in North America on dry stony soil.

■ **History** The name comes from the ancient Greek physician Dioscorides who called all thistle plants *silybon*. *Marianum* was added later from the legend that the Virgin Mary's milk ran down the leaves causing their white variegation. By association it was used to encourage the flow of milk for nursing mothers.

■ **Characteristics** A completely hardy annual or biennial with prickly, pinnately lobed, white-veined and marbled spiny leaves. The thistle-like flowerheads appear in midsummer and are followed by black seeds. Height up to 4 feet, spread 24 inches.

■ **Growing tips** Sow seed in a sunny well-drained place and thin to 2 feet apart. They do not like having wet feet. To retain the foliage effect for longer, cut off the flowering stems.

■ **How to use** Medicinally, the seeds can help liver function. Milk thistle not only protects liver function, it can actually enhance and stimulate new liver cells to form. This action is of great interest in the treatment of liver cirrhosis and hepatitis, both potentially fatal conditions.

■ **How to take** Take 20 drops of liquid tincture twice daily or a standardized capsule of powdered milk thistle extract.

comfrey *Symphytum officinale*

In medieval times, comfrey had so many self-help medicinal applications it was looked upon as a cure-all. Comfrey is a native of Asia and Europe and it grows freely in temperate regions of North America. Indeed, it can attain rampant weed proportions if not contained.

History The roots and leaves were applied to swellings, sprains, bruises, and cuts, and as a poultice to abscesses, boils, and stings.

Characteristics The plant, a perennial, has dull, dark green, hairy leaves, which may be up to 8 inches long; they have no aroma, and are not in the least attractive. The bell-shaped flowers are cream flashed with red, and hang in clusters. The roots are thick, tapering, and persistently multiply. The plant grows to 36 inches tall, with a spread of 18 inches.

Growing tips One plant, or even a piece of root with or without a shoot is all that is needed to establish a patch of comfrey. Root divisions may be made at any time except during midwinter. Plant the piece of root in deep, moist soil; the roots can penetrate to at least 36 inches. With a high concentration of potash, the plants provide a natural source of nutrition for the soil.

How to use Comfrey is related to borage, and can be used in similar ways. The young leaves can be cooked and eaten as spinach.

Comfrey is a potent herb for clearing problem skins. Use the root to make decoctions, the leaves to make healing antiseptic infusions, face packs and steamers, creams and lotions. Medicinally, comfrey is probably one of the best know therapeutic herbs. Comfrey creams are a very effective remedy for healing wounds, eczema, psoriasis, haemorrhoids, and skin ulcers.

How to take As a cream, use locally as required.

Warning Do not take internally.

thyme *Thymus vulgaris*

Thyme is a sun-loving herb, at its aromatic best when growing wild on the sun-baked hills around the Mediterranean. Different species include lemon thyme **Thymus citriodorus**, *and caraway thyme* **T. herba-barona**. *The English wild thyme referred to by Shakespeare is* **T. drucei**.

■ **History** Thyme is one of the oldest recorded culinary herbs, probably in use well before the time of the ancient Greeks. The Romans took it to Britain as part of their culinary armory. In his herbal, Nicholas Culpeper credits it with a singular usefulness: "An infusion of the leaves," he has written, "removes the headache occasioned by inebriation."

■ **Characteristics** Thyme is a low-growing sub-shrub that can become untidily woody and straggly. It can reach a height and spread of about 8 inches. The leaves are very small, only about ¼ inch long; according to type, they may be green, gray-green, yellow, or variegated. The flowers, which cover the plant from early summer, are borne in clusters at the tips of the shoots.

■ **Growing tips** Thyme thrives on poor, stony soil as long as it is planted in full sun. It is grown from tip cuttings taken in summer, or by layering stems. It will need protection in a harsh winter. Wild creeping thyme, *T. pulegioides*, is planted as a flowering lawn.

■ **How to use** Thyme is traditionally used with parsley in stuffings for chicken and pork, and, with the addition of a bay leaf, in *bouquet garni* . It is especially good with oil in marinades, and with vegetables such as zucchini, eggplant, and tomatoes. Use the leaves to make astringent infusions and strongly perfumed herbal oils for soothing body rubs. Thyme is also antiseptic and deodorizing and has anti-dandruff and conditioning properties for the hair. The pungent essential oil is useful in combating acne.

Taken internally for coughs and colds or more serious problems such as bronchitis and asthma, the mucous clearing ability of thyme makes it the appropriate remedy where catarrh is a problem.

■ **How to take** Take 20 drops of liquid tincture twice a day.

fenugreek *Trigonella foenum-graecum*

Fenugreek has been grown around the Mediterranean region since ancient times. The sprouting seeds may be eaten as a spicy salad; the fully developed leaves and the lightly roasted seeds are used as a spice, principally in curries. The ground seeds, containing coumarin, are a major ingredient in commercially prepared curry powders.

- **History** The plant is a native of western Asia and has been widely grown in countries bordering the Mediterranean, particularly in Egypt. Its cultivation in northern Europe was principally intended for forage, to mix with hay crops.

- **Characteristics** The plant, a half-hardy annual, grows to a height of 24 inches, with a spread of 8 inches. The leaves are trefoil, rather like clover, and the flowers, which appear in late spring, are cream or pale yellow and pealike. The seed is compact and pale brown. Light roasting brings out the full flavor.

- **Growing tips** The seed may be sown indoors in mid-spring or outdoors in warm soil in late spring. Fenugreek likes a good, well-drained soil and a position in full sun, which is essential if the seed is to set.

- **How to use** The sprouted seeds are good as a salad, tossed in a vinaigrette dressing. The roasted seeds are used in Middle Eastern variations of *halva*, a rich

sweetmeat, as well as in curries. It is believed to impart a sweet odor to the body when a tea made from the seeds is drunk regularly. A poultice made from the crushed seeds, boiled for 10 minutes and applied to afflicted areas, will improve a poor skin, ridding it of persistent spots and pustules.

Medicinally, fenugreek has the ability to reduce muscular spasm. This has made it the herb of choice in menstrual and labour pains. The traditional use of this herb has been in the treatment of non-insulin dependent diabetes, inflammation of the stomach, digestive problems, and painful periods.

An external application can be of help in arthritis.

- **How to take** Take 20 drops of a liquid tincture twice daily.

valerian *Valeriana officinalis*

Not to be confused with the red valerian, **Centranthus ruber,**
Valeriana officinalis *was known in medieval Europe as*
"All Heal." It is a tall airy plant for a damp spot in the garden
in either sun or shade. Valerian is native to Europe and
temperate regions of Asia and has naturalized in America.
It grows in damp grassland, meadows, and by water..

History The name probably comes from the Latin *valere*, to be well. Valerian was recommended by Hippocrates and was recorded in the Anglo-Saxon "leech books" in the eleventh century. It was widely grown in the monasteries as a spice and perfume. The smell of the root is rather like sweaty leather, and is very attractive to cats and rats. It is said that the Pied Piper of Hamelin carried a piece in his pocket! It was used in World War I to treat shell-shock.

Characteristics Valerian is a hardy perennial with hollow stems that grow up to 4 feet. It has bright green, toothed, pinnate leaves and pale pink or white flowers in terminal clusters from early to midsummer.

Growing tips Propagate by seed sown indoors or outside in spring. Plant in sun or shade in any soil but where the plants will have a cool root run.

How to use A good handful of the grated dried valerian root and double the quantity of dried chamomile flowers added to the water will provide a sleep-inducing bath.

Medicinally, the root is used to help induce sleep and relaxation, allowing the body to divert its healing powers to where they are most needed.

The traditional remedial uses for this herb include: hysteria, cramps, indigestion, high blood pressure, painful periods, palpitations, and insomnia.

How to take Try 25 to 30 drops of the liquid extract before bed as a sleeping remedy.

ginger *Zingiber officinale*

Gingerroot has been enjoyed for centuries for flavoring. It is an essential ingredient in Eastern dishes and has a long-standing importance as a medicinal herb. Originally from south-east Asia, ginger is a tropical herb which grows in Australia, Africa, South America, the West Indies, Florida, China, and Japan.

- **History** The name comes from the Sanskrit *singabera*. Ginger was described by the Chinese in the Han dynasty (AD 25–200) as the "universal medicine." The Greeks and the Romans imported it from the East in AD 200. From the fourteenth century it was the second most important spice after pepper in England. In the sixteenth century it was taken by the Spanish from the East Indies to the Americas.

- **Characteristics** Ginger is a deciduous perennial with thick branching rhizomes, stout upright stems, and large lanceolate, bright green leaves about 8 inches long. The flowers are yellow-green with a deep purple lip and yellow spots and markings. They are followed by fleshy capsules. Height to 9 feet.

- **Growing tips** Ginger is a tender crop, usually treated as an annual. In hot countries it can be planted outside in sun or part-shade. In cooler zones it can be grown outside in summer as a foliage plant but needs shelter when frost threatens. It is propagated by division (it divides easily) in spring and should be re-potted each year.

- **How to use** The dried grated root can be made into a strong infusion to act as a light coloring agent for red hair. Medicinally, ginger can reduce nausea and motion sickness and has become a popular herb for the treatment of morning sickness associated with pregnancy. Its safety in recommended doses is good, but excessive intake can prove dangerous. Ginger has been used as a traditional treatment for skin irritations, both externally applied and by internal dosage. For colds and flu, ginger extract has a warming effect and can boost the immune response to infection.

- **How to take** 25 drops of liquid tincture twice daily, or take a capsule of concentrated dried ginger root. As a tea, crush a slice of fresh ginger and add to the infusion.

Herbs for cosmetics

IN TODAY'S SOCIETY, BOTH MEN AND WOMEN ARE BECOMING increasingly aware that natural and organic substances can be used more effectively in the care of their skin, body, and hair than commercial and chemical preparations. Whether or not you believe that commercial cosmetics are harmful, it is certain that natural ingredients can do nothing but good. Herbs, vegetables, fruit, nuts, pulses, and grains were once the only means by which healing ointments and creams were made, for they contain valuable oils, pectin, minerals, and vitamins and a complexity of constituents which are today being examined more closely and tested by the pharmaceutical and cosmetic industries.

Learning of the curative potential of natural ingredients allows you the freedom to choose herbs to suit the needs of your skin and to enjoy pampering yourself with healing baths and soothing body balms. Natural deodorants and breath fresheners are free of chemicals which might have adverse effects when used over a period of time, while hair preparations are gentle and effective and, of course, chemical free.

Stress has a deleterious effect on our health and looks and while diet and a healthy lifestyle play a major part in improving overall health, one of the most powerful and evocative tension busters is perfume, which can be applied to indulge the senses or boost the morale.

Day creams and lotions are the basis of a beauty routine but once you know your skin's peculiarities you can supplement them with the extra luxury or emergency aid that herbal cosmetics provide. From fragrant face creams to simple hand and foot lotions, these cosmetics cost so little to prepare and are creative, therapeutic, and gratifying to make. ■

Herbal preparations and techniques

The following ingredients and techniques demonstrate the simple and rewarding methods for making cosmetic preparations at home for your own body. This practical pursuit has a special therapeutic value.

Infusions are strong teas made with herbs or flowers and which are often used as therapeutic elements in making skin creams and lotions. Infusions also make excellent facial rinses. The proportions given below can be altered to give a stronger or weaker infusion. They are best made in a china teapot or bowl but never aluminum as it will give a tainted result. Keep for no longer than three days in the refrigerator.

1 *Place 4 oz fresh, or 2 oz dried, herbs or flowers in the teapot and pour over 2 cups of boiling purified or mineral water. Cover and leave to stand for three hours.*

2 *To strain, slowly pour the infusion through a filter paper placed within a funnel.*

Flower waters are used in cosmetics in the same way as infusions but they are stronger, sweeter smelling and usually made specifically for use in toilet waters and perfumes. The method for making them is the same as for infusions but the liquid is left to stand overnight. Elderflower, rose, and lavender make particularly good scented waters.

Herb and flower essences are made with essential oils and are used to add extra strength to homemade herb and floral waters which may not always have the depth of perfume required, or for use as an economical toilet water. To make, take one tablespoon of essential herb or flower oil and pour

into a bottle with two cups of pure alcohol, vodka or gin. Shake well and leave for several days before using.

To dilute for use, add one tablespoon of the essence to two cups of purified water.

Essential oils are obtained from the odoriferous substances of flower, bark, seed, grain, root, resin and peel of wild and cultivated plants. They are extremely concentrated, and should be used sparingly and with close attention to the labeled instructions. The processes of manufacture involve the use of vast quantities of raw materials that makes them expensive. Nevertheless, it is advisable to purchase a good quality product because cheaper oils are often diluted or adulterated and will not have the same result.

Herb or flower oils do not have the same concentration as commercially produced essential oils. However they can be used as a fragrant and therapeutic substitute for olive oil in many recipes for body lotions and skin preparations, and are particularly useful in massage oils.

The quantities used are 2 oz fresh, or 1 oz dried, herbs or flower petals to 2 cups olive oil and 1 tablespoon pure alcohol, vodka or gin, or cider vinegar.

1 *Crush the petals or herbs with a little of the oil in a mortar with a pestle.*

2 *Transfer to a large glass jar and add the remainder of the oil and the alcohol or vinegar. Seal tightly and shake well.*

3 *Stand the jar on an inverted saucer in a deep saucepan and submerge it to the level of the oil in warm water. Heat very gently on a low, regulated heat for an hour. Repeat daily.*

4 *After two weeks, strain the mixture through a nylon strainer, pressing it well. Repeat the process until the oil has a good, warm fragrance. Twice is usually enough. Bottle and seal.*

Herb and flower vinegars are therapeutic. Rose and elderflower will relieve the tension of headaches, while lavender eases aching muscles, heat, and inflammation. When used as skin toners or added to bath water, these gentle vinegars will also restore the pH balance of the skin. They are excellent when used in astringents and give a conditioning gloss in a final hair rinse. The best vinegars to add to the bath are those containing a wide selection of herbs and flowers. The proportions are usually 2 oz of fresh, or 1 oz of dried, herbs or petals to 2 cups of cider vinegar. Fresh ingredients are used as they are, but dried herbs and petals should be pounded in a little vinegar in a pestle and mortar before using.

1 *Warm a large, plastic-lidded glass jar and spoon in the herbs or petals.*

2 *Heat the vinegar until just hot and pour it slowly into the jar.*

3 *Seal tightly, shake well and leave on a sunny windowsill for one month, shaking daily.*

4 *Strain, re-bottle and seal.*

Herbal steamers. The use of a herbal steamer is a most effective way to deep cleanse the skin. Steaming relaxes the skin, encouraging it to release impurities, while at the same time stimulating the release of nutrients to the skin cells and the removal of toxins. The herbs used have a variety of therapeutic as well as cosmetic effects. They heal, soothe, cleanse, stimulate, and tighten the skin. Some are effective in drawing pimples to a head, others for cleansing pores but at all times they are calmative and relaxing.

It is advisable not to use a steamer if your skin is very dry or sensitive or has broken or thread veins, as heat may cause excess irritation. It may also exacerbate skins suffering from acne and can lead to the spread of infection. Steamers should never be used by people suffering from asthma or other breathing difficulties.

1 *Cover the hair and thoroughly cleanse the face.*

2 *Place two handfuls of the appropriate herb or herbs in a large bowl and cover them with 3½ cups of boiling water.*

3 *Hold your face about 12 inches above the water and cover both head and bowl with a thick towel to prevent the steam escaping. Stay under the towel for ten minutes.*

4 *Simply blot dry and then refresh with a toner, an infusion, or tepid water, followed by a moisturizer. Avoid a cold atmosphere until your skin has cooled down.*

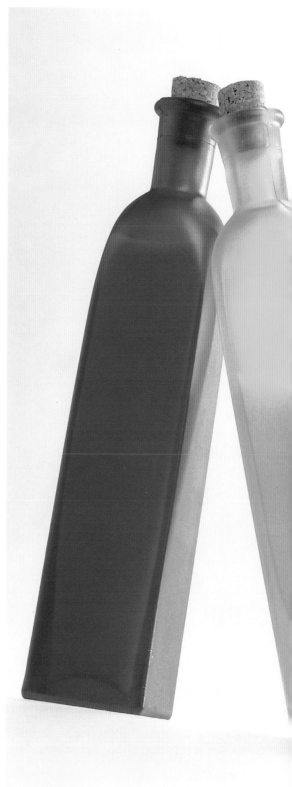

Face masks and poultices. All skins benefit from the regular application of a cleansing and stimulating face mask which not only improves the circulation but also encourages a healthy flow of nutrients to the skin, thereby enhancing its tone, texture, and color. Some face masks are healing with a deep drawing action to remove impurities from problem skins, while others rejuvenate tired or mature skins. Others may moisturize, bleach or tone.

Use fresh herbs when possible because they make a more substantial poultice and can be used without the addition of thickening agents such as oatmeal, fuller's earth or kaolin. However, many of these additions are of value in their own right. Oatmeal is whitening and softening, fuller's earth is cleansing and stimulating, and kaolin has deep drawing powers. Dried herbs make perfectly adequate face masks but can only be used when warm as they have to be reconstituted in hot water before becoming effective. Infusions of both fresh and dried herbs can be used in conjunction with thickening agents.

Before embarking on a home facial, cover the hair well and ensure that the skin has been cleansed. When applying any face mask always avoid the delicate area around the eyes and mouth. Allow the mask to work for a prescribed time and then remove with tepid water.

How to make hot masks

1 *Chop the chosen herb or herbs and soak in enough hot water to cover until soft. Squeeze out excess water but leave enough to ensure a good "mash."*

2 *Apply the herbs thickly to the skin either as they are, or for both therapeutic benefit and ease of application, combine with a thickening agent.*

How to make cold masks

1 *Take a good handful of clean fresh leaves and petals and pulverize them in a blender with a little purified water.*

2 *Apply to the face as they are, or combined with a thickening agent.*

Bain marie Recipes to make your own creams, ointments, balms and skin toners require the use of a *bain marie*. Waxes and oils are melted in a china bowl, standing in a pan containing several inches of water. The water is brought to a simmer to allow the contents of the bowl to be heated without direct contact with the heat, which might otherwise result in burning or sticking. A separate bowl is used in the same way to warm liquids, when two sets of ingredients need to be brought to the same temperature simultaneously.

1 *Place a bowl over a pan of simmering water.*

Containers and storage It is not a good idea to make too much of any one preparation. The amounts given in the recipes allow for sufficient to discover whether they are suitable for your skin and to ensure that they do not lose their freshness over long storage. All containers must be scrupulously cleansed and dried before use, and must have plastic lids, corks or glass stoppers. Tight seals on containers are important when traveling, particularly to hot countries, as many of the ingredients have a low melt point. Although the addition of tincture of benzoin will act as a preservative, particularly when an infusion is used, it is always best to refrigerate natural products.

2 *Waxes, which should be grated or finely chopped, are placed in one bowl with any oils (except essential oils) to slowly dissolve in the gentle heat. Liquids, such as water, infusions, etc. are placed in a separate bowl and heated in the same manner.*

Allergic reactions It is very rare to have an allergic reaction (usually recognized by irritation, a small rash or tiny pimples) to a recognized and well used herb, especially if you are using one that is suitable to your skin type. Some waxes and oils may cause an adverse reaction as do some essential oils. If you are in doubt then follow this simple test.

1 *Take a small strip of non-allergenic sticking plaster and place a little of the substance – herb, wax, oil etc. on the gauze padding.*

2 *Tape it firmly to the soft skin on the inner arm above the elbow joint.*

3 *Leave the plaster on for 24 hours by which time the skin will have reacted with a rash or small pimples.*

Glossary of ingredients

- **fuller's earth** – a highly absorbent clay used at one time in the processing of woolen cloth as it absorbs fat. For the same reason it removes excess grease from the skin and alleviates problems caused by clogged pores. Available from good pharmacies or herbalists.

- **brewer's yeast** – a yeast used in brewing, and which can be used to stimulate the skin and draw out pimples. Rich in vitamin B, minerals, and proteins, it is particularly useful in cleansing masks for oily skins. Available from health food stores, good pharmacies, and herbalists.

- **kaolin** – a fine white to yellowish or gray clay used as an absorbent in medicine and cosmetics. Kaolin poultice was once and is still used for its unique properties as a drawing agent for serious skin abscesses. It is also a good binding agent. Available from good pharmacies and herbalists.

- **borax** – hydrated sodium borate is used to soften water and acts as an emulsifier. It also has detergent qualities. Available from good pharmacies.

- **beeswax** – a yellowish wax found in natural honeycomb. It is a nourishing emollient with a high melting point. When mixed with borax it becomes an emulsifier which prevents the separation of water and oils when combined in a recipe. Available from most health food stores, craft shops, herbalists, and honey suppliers.

- **almond oil** – the oil obtained from almonds. It is rich, pure and used extensively in beauty preparations and remedies.

- **aloe vera juice** – obtained directly from the plant (see The Herb Directory, page 53). It is available from health food stores, herbalists, and good pharmacies. Always look for the International Aloe Science Council (IASC) Seal of Approval to ensure a pure product.

- **glycerin** – a thick, colorless liquid which can be dissolved in water. Because it is oily it is often used in remedies for dry skin and in bath products. Available from pharmacies.

- **witch hazel** – an astringent, antiseptic distillation of the bark of witch hazel, used widely in pharmaceutical and cosmetic preparations. Available from pharmacies.

- **turkey red oil** – a specially treated castor oil that will dissolve in bath water and is therefore readily absorbed into the skin. Available from specialist suppliers (see page 219)

- **anhydrous lanolin** – a rich, sticky fat obtained from sheep's wool. It easier to use than hydrous lanolin (which contains water and is not as suitable for skin preparations) Available from specialist suppliers (see page 219).

- **avocado oil** – rich in vitamins, minerals, and lecithin. Pure and very moisturizing. Available from specialist suppliers (see page 219).

- **cocoa butter** – a rich wax obtained from the cocoa bean, and used as an emollient. Available from specialist suppliers (see page 219).

- **olive oil soap** – soap made using olive oil. While the soap is quite coarse, it is pure. Available from health food stores and herbalists.

- **tincture of benzoin** – an aromatic, natural resin which is used as a preservative in skin preparations. It is also known as Gum Benjamin or Friar's Balsam. Available from specialist suppliers (see page 219).

- **dried herbs** – almost every variety can be obtained from good herbalists and by mail order from specialist suppliers.

yarrow

Yarrow is the most efficient healing herb for oily and problem skins. It has both drawing and astringent properties and is most frequently used in poultices and steamers and in conjunction with those herbs which will best deep cleanse skins prone to acne and enlarged pores: comfrey; dandelion, catmint; sage; tansy and feverfew. Yarrow infusion, or the juice extracted from the leaves, will reduce dilated facial veins.

yarrow and nettle cleanser

A mildly astringent healing lotion suitable for sallow greasy problem skins which is also softening and cooling. The same recipe, more appropriate for dry and normal skins, can be used by substituting elderflower, tansy or feverfew.

> **1 cup buttermilk**
> **2 oz fresh yarrow and nettle tops**

Simmer the herbs with the buttermilk for 20 minutes. Remove from the heat, cover and infuse for two hours. Strain, bottle and keep refrigerated.

cleansing face pack

An excellent facial for skins which are greasy and prone to pimples and mild acne. The same treatment can be used with fennel or lady's mantle for mature skins. It is not suitable for dry or sensitive skins and should not be used more than once a week.

> **4 tablespoons fuller's earth**
> **2 teaspoons plain yogurt**
> **2 teaspoons strong yarrow infusion**
> **1 teaspoon clear honey**

Blend the ingredients into a smooth paste and apply the mixture to the face avoiding delicate skin around the eyes and mouth. Leave in place for about 20 minutes or until quite dry, refraining from the temptation to move a muscle. Rinse off with a clean cloth and plenty of tepid water. Finish with a herbal infusion.

aloe vera

The gel obtained from the fleshy leaves of aloe vera has a 90 percent water content which allows valuable constituents, minerals, and vitamins to be carried deep into the dermal layers of the skin. It is not a sun block but applied as a skin lotion the gel can, in time, eradicate small scars and age spots and smooth out facial lines. Aloe vera is non-allergenic and antibiotic and is therefore suitable for sensitive skins.

after-sun moisturizer

The added vitamins ensure that there is extra nourishment for skins suffering from climatic damage.

- 1 tablespoon beeswax
- 6 tablespoons almond oil
- 3 tablespoons distilled or mineral water
- ¼ teaspoon borax
- 2 tablespoons aloe vera juice
- 1 capsule each vitamins A and E

Using a *bain marie*, heat the grated wax and oil in one bowl. Dissolve the borax in the water in another, and the aloe vera juice in a third bowl. When the beeswax has melted remove all the bowls from the heat. Mix the borax solution with the aloe vera juice and pour slowly into the beeswax mixture, beating constantly until it begins to cool. Prick the vitamin capsules and stir the contents into the cream. Continue beating until quite cold. Pot into sterile jars and seal. Keep refrigerated.

massage lotion

A few drops of elderflower, eucalyptus, lavender or thyme can be added to give therapeutic benefits and perfume. This ointment should be kept refrigerated as it melts at room temperature.

- 4 tablespoons cocoa butter
- 4 tablespoons coconut oil
- ½ cup olive oil
- 1 cup aloe vera juice
- 9 drops of essential oil (optional)

Heat the cocoa butter and oils in a *bain marie* and the aloe vera gel in another. When the cocoa butter has melted remove both pans from the heat and slowly pour the aloe vera juice into the wax mixture. Continue beating until cool when the essential oil should be added if used. Beat until cold. Pot into sterilized jars and seal.

lemon verbena

The flowers, leaves, and stem of lemon verbena are delicately fragrant with a slightly lemony, vanilla smell that is very relaxing and refreshing when added to bath water, or dried and used in potpourri. An infusion made from leaves and flowers will improve the sheen and add highlights to dark hair. The essential oil of verbena is frequently used in healing treatments for acne because of its antiseptic properties. It can be added to cleansers and moisturizing creams but should always be used with caution on sensitive skins.

lemon verbena sun oil

A simple moisturizing oil. The perfume acts as both an insect deterrent and perfume. Citronella, lemongrass or essential oil of lavender can be substituted for verbena.

- ½ cup sesame oil
- 5 tablespoons cider vinegar
- 1 teaspoon iodine
- 6 drops essential oil of verbena

Shake the ingredients together in a bottle, keep well sealed and shake before use.

floral after-bath cologne

This lovely cologne relies upon the fragrance of fresh flowers for its depth of perfume.

- 1 large cup fresh petals: rose geranium, jasmine, roses, etc
- 1 large cup pure alcohol or vodka
- 3 cups water
- 6 tablespoons dried ground orange and lemon peel
- ¼ teaspoon ground cloves
- 2 tablespoons crushed lemon verbena
- 2 tablespoons dried crushed mint

Put the petals into a jar with the alcohol. Seal tightly and leave for one week. Next, boil the water and make an infusion of the dried peel, herbs, and cloves. Cover and leave to stand for 24 hours. Strain both liquids through a fine strainer and then combine. Pour into a jar with a tight stopper and shake well.

chamomile

Chamomile flowers are used to make mildly astringent infusions for tonic cleansers and conditioners that are suitable for dry and normal skins. Healing and gentle, it can be used in face packs and steamers to treat problem skins of all types and ages. A cool compress of chamomile flowers simmered in milk and strained will heal eczema, flaking patches, rashes and thread veins and reduce wrinkles. Traditionally, chamomile is used as a hair colorant which conditions and lightens fair hair.

chamomile skin tonic

An excellent cleanser which conditions oily skin.

2 oz fresh, or 1oz dried, chamomile flowers
2 oz fresh rosemary tops
2 cups boiling hot water
1 teaspoon fresh lemon juice

Make an infusion and leave overnight to cool. Strain, add the lemon juice and refrigerate.

chamomile lightening paste

The more often this paste is used the better the tone.

1 cup strong chamomile infusion
8 tablespoons kaolin powder
1 egg yolk

Mix the ingredients together and apply to the hair working from the roots. Cover and leave between 20 minutes and 1 hour depending upon the porosity of the hair. Rinse off with warm water. Use a nourishing shampoo and conditioner.

marigold

Infusions of marigold petals are suitable for treating eczema, acne, blemishes, scarring, blisters and burns. It is especially good for sensitive skin and is used professionally to treat dermatological problems. The herbal oil and infusion will gently condition ageing skin, helping to reduce wrinkles while marigold petals, pounded with warm wheat germ oil, is a healing remedy for small blemishes, scars and thread veins. The infusion also makes a conditioning hair rinse, giving golden red hues to fair hair.

marigold and yogurt paste

A healing and cleansing scrub, suitable for blemished and ageing skins which have been exposed to sun and wind.

- 1 tablespoon fresh or ½ tablespoon dried marigold petals
- 1 teaspoon wheat germ oil
- 1 teaspoon clear honey
- 1 tablespoon plain yogurt
- 1 teaspoon fresh lemon juice

Pound the petals with the oil and honey. Add the yogurt and lemon juice and leave to stand for a few minutes before applying.

marigold astringent

A gently healing astringent tonic suitable for oily skin.

- ½ cup marigold infusion
- 1 tablespoon witch hazel

Combine the two liquids in a bottle. Shake well and keep refrigerated.

marigold cream

A light, healing, and nourishing cream suitable for mature, blemished, or climatically damaged skin.

- 2 tablespoons beeswax
- 2 tablespoons anhydrous lanolin
- 6 tablespoons almond oil
- 1 teaspoon wheat germ oil
- 6 tablespoons marigold infusion
- ½ teaspoon borax
- 2 drops tincture of benzoin

Heat the waxes and the oils together in one *bain marie* and the marigold infusion in another. Dissolve the borax in the infusion and remove both bowls from the heat. Beat the infusion slowly into the oils with the benzoin. Continue beating until thick and cool. Pot and seal in sterile jar. Keep refrigerated.

marigold jelly

A superb remedy for sore or sunburnt hands, rough and cracked skin, and hangnails.

- Petals from 6 fresh marigold heads
- 1 large jar of petroleum jelly

Melt the jelly in a *bain marie*, add the petals and simmer gently for several hours. Strain the mixture and pot.

fennel

The feathery leaves of fennel have a particularly beneficial effect on mature skins and are healing and regenerative in steamers. Mature skins need soothing face packs which will remove impurities without stressing the skin and fennel has this quality. An infusion makes a mild toner which can be used frequently on the skin to help eradicate fine lines. It is also gentle enough to be used as an eye lotion.

fennel eye lotion

This infusion will help bring a sparkle to the eyes.

1 oz fennel seeds
2 cups purified water

Simmer the seeds in the water for 20 minutes. Strain, cool and bottle. Applied to the eyes on lint pads.

fennel and olive oil facial

A wonderful refining mask for all skin types. It is also a very effective moisturizer after sunbathing.

4 tablespoons olive oil
2 tablespoons fennel infusion

Warm the olive oil in a *bain marie* and soak enough cotton wool in it to cover the face. Place protective gauze pads over the eyes and apply the oily compresses over the face. Leave until cold, blot dry with tissues and follow with a warm compress of the fennel infusion. Finish by splashing with cold water.

orris root

Orris powder has been used for many centuries as the base of face and talcum powders, and as an ingredient in bath salts, shampoos, toothpaste, and potpourri. It has a delicious fragrance similar to violet. Oil or otto of orris is extracted by distillation of the roots under steam. Less fleeting than violet, the perfume is used extensively in the perfumer's art.

poudre a la mousseline

This and similar powders are excellent to use as dusting or talcum powders. Other herbs and spices can be used. Vanilla powder and lavender flowers are particularly appropriate. A few drops of essential oil can also be added to the powder but care is needed to blend without sticking.

9 oz finely ground orris root

5 oz cornstarch

2 oz fine rice flour

6 oz powdered coriander

2 oz powdered cloves

1 oz powdered cinnamon or cassia bark

1 oz powdered sandalwood

Make sure that all these ingredients are powdered as finely as possible.

Mix thoroughly and put into a large airtight jar. Shake well and frequently.

herbal bath salts

A handful of these salts neutralizes the acids in the skin and softens the hardest water.

5 oz bicarbonate of soda

3 oz orris root

a few drops of essential oils such as rose geranium, sandalwood, pine, lavender

Mix the ingredients together and then pound in a pestle and mortar until the perfumes are well integrated. Store in an airtight jar.

lavender

Lavender is probably best known in traditional toilet water but it has been used as an antiseptic and perfumed addition to skin care preparations throughout the centuries, as well as being an essential element in home cleaning products because of its ability to deter infection and insects. The pungent and healing flowers and leaves make excellent infusions, water, and herbal oils which can be added to all preparations for skin and hair care. Steamers and face masks made with lavender are beneficial for skins which have blemishes. The slightly astringent infusion in creams and lotions helps problem skin and to tone and condition oily hair.

It relaxes tired muscles in a soothing bath and energizes in after-bath colognes. Cooling and refreshing it makes excellent foot care preparations and as an antidote to sunburn. The essential oil of lavender can be used in a base oil as a massage to reduce cellulite, to heal acne, and as a remedy for eczema and itchy scalp. Commercial lavender water can be used as a substitute for an infusion

lavender foot balm

A soothing, emollient balm.

- 6 tablespoons anhydrous lanolin
- 3 tablespoons almond oil
- 3 tablespoons glycerin
- 6 drops essential oil of lavender

Melt the lanolin in a *bain marie* then beat in the oil and glycerin. Beat the mixture until it is nearly cold and then add the lavender oil. Lavender oil or flowers added to a footbath are very therapeutic, and so is lavender herbal oil made with olive oil. Rose geranium and marigold also relieve and soothe tired feet.

lavender body lotion

A soothing, healing lotion, perfect for a massage.

- 4 tablespoons almond oil
- ½ cup lavender infusion or rose water
- 1 teaspoon borax
- 8 drops essential oil of lavender

Warm the almond oil in one *bain marie* and the infusion or rose water and borax in another. Remove both bowls from the heat and pour the liquid slowly into the oil, beating continuously until cool and no longer separating. Add the essential oil and beat until cold. Pour into a sterile bottle. Seal and shake well before using.

Essential oils of violet, rose, rose geranium, ylang-ylang, and neroli can be substituted for the lavender.

lavender cleansing cream

A simple, light cleansing cream suitable for all skin types and which can be adapted to suit a variety of skin conditions by substituting any flower water or herbal infusion for lavender.

- ½ oz white wax or beeswax
- 6 tablespoons almond oil
- 5 tablespoons lavender infusion or water
- ¼ teaspoon borax
- 4 drops essential oil
- 2 drops tincture of benzoin

Melt the wax and oil together in a *bain marie* and warm the lavender water with the borax in another. Remove both bowls from the heat and pour the water slowly into the waxes beating continuously. When cool add the essential oils and the benzoin and continue to beat until cold. Pot and seal in sterile jars.

lavender deodorant

A fragrant, healthy, and long-lasting deodorant.

- 3 drops essential oil of lavender
- 1 tablespoon sugar
- 2 cups purified water

Shake the ingredients together in a bottle and leave for two weeks. Decant into an atomizer and shake before using

mint

Garden mint will give subtle fragrance and therapeutic benefit when used for infusions in skin preparations. An infusion or extraction of the juice can be used for quick toners and "pep-up" remedies for a tired complexion. The juice from crushed mint will help alleviate and diminish the dark shadows under the eyes. Sometimes this is hereditary but more often the result of tiredness or stress.

A lavender, mint, and rosemary bath bag with one tablespoon of fine oatmeal added and tied under the hot running water makes sure of a silky smooth, relaxing bath.

mint and rosemary mouthwash

A pleasantly antiseptic rinse.

1 teaspoon fresh chopped mint
1 teaspoon fresh or ½ teaspoon dried rosemary
2 cups purified water
½ teaspoon tincture of myrrh

Make an infusion with the water and herbs.
Allow to cool before filtering and adding the myrrh.
Keep refrigerated.

mint and parsley moisturizing milk

A light and antiseptic moisturizer for those oily skins which are prone to blemishes and for which a heavy cream is unsuitable.

3 tablespoons chopped fresh mint
3 tablespoons chopped fresh parsley
1 cup milk

Bring the milk to simmering point and pour into a warmed sterile jar with the herbs. Seal, shake and leave for 12 hours. Strain and bottle. Keep refrigerated and use both morning and night.

mint skin cream

A nourishing cream for most skin types. Sage infusion may be substituted for mint as sage is suitable for dull and sallow skins.

1 teaspoon beeswax
6 tablespoons coconut oil
4 tablespoons olive oil
2 tablespoons almond oil
½ cup strong mint infusion
1 teaspoon borax
2 drops tincture of benzoin

Melt the wax and oils together in a *bain marie*.
Dissolve the borax in the infusion and heat it in another bowl. Remove both bowls from the heat and slowly pour the liquid into the oils beating constantly until cool. Add the tincture of benzoin and beat until cold. Pot into sterile jars and seal.

mint eye lotion

After applying this lotion it is a good idea to soak cotton wool pads with a fine oil–apricot or almond– and place them on your eyes while taking a rest.

A good handful of very fresh mint leaves

Turn on the blender and feed the leaves onto the blades until they are thoroughly pulverized. Strain through a nylon strainer or squeeze through a cheesecloth bag. Use the juice to relieve tiredness in your eyes. Keep any surplus refrigerated.

peppermint

Peppermint contains menthol which not only makes it invaluable in preparations for the mouth but also in toners and body massage lotions. Used in astringents it will reduce open pores and tighten and cleanse oily skin. Added to steamers and face packs, it is kind to sallow and greasy skins as it draws and stimulates. It is most effective on the dull and oily areas around the mouth and nose which are a main concern to people with combination skins.

Peppermint oil extract (not essential oil) can be added to a variety of face masks. Chopped peppermint or peppermint extract added to a warm footbath is extremely soothing. Essential oil of peppermint should not be used without the guidance of a professional aromatherapist.

peppermint lemon toner

A stimulating and antiseptic tonic suitable for oily skins.

- 4 tablespoons fresh lemon juice
- ½ teaspoon peppermint oil or extract
- ½ cup witch hazel
- 2 tablespoons gin or vodka

Combine all the ingredients and leave to stand for 24 hours. Pour into a sterile bottle and shake before using.

peppermint face mask

A zingy mask which is a good remedy for sallow or blemished skin. Not recommended for sensitive skin.

- 4 oz brewer's yeast
- 1 tablespoon witch hazel
- 4 drops of peppermint extract

Mix the ingredients to a paste. Apply to the face and leave on for 30 minutes. Wash off with warm water followed by a herbal infusion or a little lemon juice.

A word of warning: brewer's yeast brings pimples to a head, so do not use just prior a special occasion.

energizing body oil

An easily absorbed oil that will help start the day well. The perfume can be changed to suit your mood by using any fragrant essential oil.

- 4 tablespoons sunflower oil
- 4 tablespoons olive oil
- 2 tablespoons almond oil
- 2 tablespoons mineral oil
- 1 teaspoon peppermint extract

Pour ingredients into a bottle and shake well.

peppermint mouthwash

This is a very quick and refreshing mouthwash which can be used daily. It not only cleanses the breath but is astringent and therapeutic.

- 1 cup distilled water
- 6 tablespoons witch hazel
- 1 teaspoon peppermint essence
- 1 strip finely pared lemon peel

Put all the ingredients in a bottle, seal and shake well. Keep in the refrigerator.

basil

Basil is a beautifully sweet-smelling herb which can be used fresh in bath water, in steamers, and in face packs for its antiseptic and relaxing properties. Basil infusions and herbal oil made with rose petals are gorgeously aromatic in creams for the face and lotions for the body. Basil and peppermint vinegar added to a bath or bath bag, tones and refreshes.

herbal after-bath oil

A basic everyday massage oil which can be altered to suit individual needs by substituting the essential oil of basil for any other appropriate oil. Pine, lavender, and eucalyptus are good for aching muscles. Rose and neroli are relaxing. Thyme and sandalwood are brisk and uplifting.

- 4 tablespoons almond oil
- 2 tablespoons each wheat germ oil and sesame oil
- 3 tablespoons each sunflower oil and olive oil
- 1 tablespoon apricot or avocado oil
- 1 teaspoon essential oil of basil

Shake all of the ingredients together in a bottle and use when needed.

rose and basil perfume

This is a variation on the ancient spice perfumes which were used to scent bath water and perfume living quarters.

- 2 cups rose water
- 1 teaspoon crushed cloves
- 1 tablespoon dried basil
- 1 shredded bay leaf
- 2 cups white wine vinegar

Combine all the ingredients in a saucepan and bring to a boil. Continue to simmer and as the liquid reduces make up the quantity with water. Cover and leave for 24 hours. Strain and bottle. Keep for one month before using.

basil and rose water lotion

A gloriously perfumed body toner which is especially effective after exercise or after a day on the beach.

- 4 tablespoons sunflower oil
- 5 tablespoons basil infusion
- 5 tablespoons rose water
- 1 teaspoon borax
- 6 drops essential oil of basil

Warm the oil in one *bain marie* and the infusion and rose water in another. When both are at the same temperature, dissolve the borax in the liquids. Remove both bowls from the heat and slowly beat the borax solution into the oil. Continue beating until the mixture combines without separating. Add the essential oil of basil, stir well and bottle. Shake well before use.

basil and lemon face mask

This is a delicious method of soothing and healing skin which feels damaged by cold weather.

- 1 handful of fresh basil leaves
- ½ avocado
- 1 teaspoon lemon juice
- 1 teaspoon clear honey

Pulverize the basil leaves in a blender. Mash the avocado flesh. Mix all the ingredients together until they are smooth. Apply to a clean face and leave for as long as possible. Rinse off the mask with tepid water.

rose geranium

The lovely, musky, sweet perfume of rose geranium oil comes from the leaves of scented pelargoniums and it is frequently used as a substitute for rose in the production of cosmetics and perfumes. Not only does the smell have a tonic effect but the herb has remarkable healing and rejuvenating properties that are invaluable in creams and lotions for mature skins. Add the essential oil to body lotions, especially in the summer, for it is also a good insect repellent and will ease stings and bites. The pungent smell of the fresh leaves and flowers will give an invigorating fragrance to a warm bath and a lasting perfume to potpourri. Essential oil of rose, which is regenerative, can be substituted for rose geranium.

sweet leaf face mask

This cosmetic remedy helps to diminish wrinkles and is very therapeutic. The petals of the Madonna lily can be used in the same way for the same effect.

1 handful of rose geranium leaves
2 cups of warm rose water

Put the leaves to soak in the rose water until they are soft. Arrange them over the face in those places most necessary. Lie down for at least 20 minutes. Tone with the rose water remaining in the cup.

healing hand cream

A rich, thick protective cream

3 tablespoons anhydrous lanolin
2 tablespoons almond oil
2 tablespoons glycerin
8 drops essential oil of rose geranium

Melt the lanolin in a bowl in a *bain marie*. Beat in the almond oil and glycerin. Remove from the heat and continue beating until the mixture cools and then add the essential oil.

rose geranium skin tonic

This is one of the most delicate and fragrant skin tonics and is most particularly suitable for older skins.

2 large handfuls rose geranium leaves
1 cup water

Simmer the rose geranium leaves in the water for five minutes. Cover and leave for 20 minutes. Strain and bottle.

Lime blossom and elderflowers make soothing and cleansing skin tonics; lilac flowers and lavender are antiseptic and fragrant; hollyhock leaves and honeysuckle flowers are softening and healing. Raspberry and blackberry leaves heal rashes and pimples.

conditioning night cream

A wonderfully emollient and rejuvenating cream which nourishes dehydrated skin.

2 tablespoons cocoa butter
2 tablespoons emulsifying wax
1 tablespoon beeswax
1 tablespoon apricot oil
1 tablespoon evening primrose oil
2 tablespoons sesame oil
1 tablespoon almond oil
1 capsule each of vitamin A, E, and D
8 drops essential oil of rose geranium

Melt the cocoa butter and the waxes in one bowl in a *bain marie* and then beat in the oils. Remove from the heat and add the contents of the capsules. Continue beating until cool and add the essential oil. Beat until cold and pot into sterile jars.

parsley

Fresh parsley is full of vitamin C and a natural cleanser. When it is applied in a face pack and most particularly when used in combination with other strong green herbs, such as mint, nettle, dandelion, or comfrey, it has a strongly tonic effect which stimulates the circulation to release impurities and improve a sallow or blotched complexion. An application of the crushed leaves or their extracted juices reduces thread veins and broken or dilated veins and bruises, while skin with open pores will benefit from an astringent parsley infusion or from cleansers and conditioners made with an infusion.

green healing mask

A good tonic mask suitable for most skins.

1 handful each of parsley and fresh spinach
1 cup water
fine ground oatmeal

Chop the parsley and spinach and boil them in the water for five minutes. Cover and leave to cool. Press through a fine nylon strainer to fully extract the juices. Mix with enough oatmeal to make a smooth paste.

If the skin is very oily but also sensitive, add plain yogurt to the mixture.

Expressed parsley juice mixed with egg white is a good tightening and boosting mask.

parsley and mint cleanser

Clears oily skin of pimples, rashes, and windburn.

½ cup chopped parsley
1 tablespoon dried mint

Infuse the two herbs in one cup of boiling water for one hour. Strain and use within three days.

parsley freckle lotion

Freckles were once considered a blemish, although not so today. However, this gentle lotion will fade them slightly and reduce the sallow end of a tan.

4 sprigs parsley
1 handful elderflower blossoms
½ cup each of milk and water

Chop the parsley and wash the elderflowers. Place them in a saucepan with the milk and water and simmer gently for five minutes. Remove from heat. Cover and leave to stand for three hours. Strain, bottle and refrigerate. Dab on freckles with cotton wool.

parsley bruise treatment

This treatment is most appropriate for fine veins and will help them diminish if used regularly. More prominent thread veins need professional treatment.

1 handful of fresh parsley
1 cup water
1 drop each essential oil of rose and marigold

Chop the parsley and boil it in the water for five minutes and then cover and leave to steep until lukewarm. Strain and add the essential oils. Leave until cold and apply using cotton wool.

rose

Rose water is the oldest beauty preparation. Used by the ancient Greeks and Romans, by the Egyptians, and in Asia, it has formed the basis of a myriad of cosmetic lotions and creams. Initially it was used for its antiseptic and healing properties but it was then found to soften the skin, reduce lines and generally have a rejuvenating effect. It smells delicious, the perfume being calmative and relaxing. Rose water is an excellent toner and rose vinegar is a good antidote to sunburn as it is exceptionally kind to sensitive skins. Rose water and glycerin is an effective balm to use on dry skin, but is not a barrier against the sun's rays. Rose petals in the bath are positively luxurious and the most strongly scented petals available are used to make water, infusions, oils, and vinegars. Essential oil of rose is monstrously expensive due to the vast amount of petals needed to produce it. A few drops added to any beauty preparation, from moisturizers to massage oil, will be kind to the skin and gentle on the soul.

rose lip balm

This balm can be used as an everyday protective salve.

- 2 teaspoon grated beeswax
- 4 teaspoons almond oil
- 1 teaspoon rose water

Melt the beeswax in a *bain marie* and beat in the oil. Warm the rose water in a separate bowl. Remove both bowls from the heat and stir the water into the oils. Pot while still warm.

rose and apricot cream

A muscle toning massage cream with a delicious fragrance. It is ideal for areas prone to flabbiness, such as the upper arms.

- 1 tablespoon anhydrous lanolin
- 1 tablespoon cocoa butter
- 2 tablespoons apricot oil
- 1 tablespoon rose water
- ½ teaspoon borax
- 4 drops of essential oil of rose

Melt the lanolin and cocoa butter together in a *bain marie* and beat in the apricot oil. Warm the rose water in a separate bowl with the borax. Beat the water slowly into the oils and continue beating until the mixture is cool. Add the rose oil and beat until cold. Pot into a sterile jar and seal.

rose moisturizer

This is an improvisation of the first cold cream made by a Greek physician for his Roman customers.

- ½ tablespoon grated beeswax
- 1 tablespoon emulsifying wax
- 6 tablespoons almond oil
- 3 tablespoons rose water
- ¼ teaspoon borax
- 4 drops of essential oil of rose

Melt the waxes with the oil in a *bain marie* and warm the rose water with the borax in another. Remove both bowls from the heat and beat the rose water slowly into the wax mixture. Continue beating until it is cool. Add the essential oil. Beat until cold, then pot into a sterile jar and seal.

rose minute mask

Pat on those areas of the face which need a bit of tightening up. Leave to dry and wash off.

- 8 tablespoons rose water
- 2 tablespoons clear honey
- 3 tablespoons witch hazel
- ½ teaspoon glycerin

Put all the ingredients into a bottle. Shake well and keep refrigerated.

rosemary

Rosemary was the ingredient at the heart of the famous Hungary water used for whitening and cleansing the skin. Rosemary herbal oil is worth the effort of making, as it is true to the perfume and well suited as a substitute for vegetable oil in massage oils and body rubs. Rosemary infusion is also used in soap, aftershave and toilet water. Add the leaves and flower tops to a bath to help relax and refresh. Dry and dark hair will benefit from rosemary infusion in a final rinse which will tone and condition hair that is thinning and lifeless, counteract hair loss, prevent dandruff and darken graying hair. Used in a base oil and massaged into the scalp it relieves irritation and dandruff.

floral skin tonic

A lovely, aromatic toner for use on skin and in the bath.

- ½ cup white wine or cider vinegar
- ½ teaspoon whole cloves
- 1 tablespoon chopped rosemary tops
- 1 tablespoon chopped lavender tops
- 1 tablespoon perfumed rose petals
- 4 cups orange blossom water

Put all the ingredients except the orange blossom water into a glass jar. Seal tightly, shake well and leave for two weeks, shaking the jar daily. Strain and repeat the process with a fresh batch of herbs and flowers. Strain again, measure and for each half-cup of liquid add four cups of orange blossom water. Bottle and seal tightly.

rosemary bath oil

Soothing bath oil will soften skin and relieve itching caused by allergy or sunburn.

- 4 tablespoons turkey red oil
- 1 tablespoon rosemary oil or
  4 drops essential oil of rosemary

Shake the ingredients together in a bottle and add one tablespoon to the bath. Any other healing and pleasant-smelling herbal oil can be used: lavender, basil, thyme.

rosemary herbal lotion

A healing, soothing cleanser for skin roughened by winter winds.

- ½ cup white wine
- 2 tablespoons rosemary leaves
- 1 tablespoon lemon balm

Simmer all the ingredients together for 10 minutes. Remove from the heat, cover and leave to cool. Strain the liquid and use on a cotton wool pad, morning and evening, to cleanse the skin.

rosemary and almond facial

A mildly abrasive scrub suitable for any skin.

- 1 tablespoon ground rosemary
- 1 tablespoon ground almonds
- 1 tablespoon finely ground oatmeal
- 2 or 3 tablespoons rose water

Mix the dried ingredients to a fine paste with the rose water. Apply to clean skin with small circular movements of the finger tips. Leave to dry and wash off with tepid water. Splash with a rosemary or rose water toner.

elder

The elder is steeped in folk lore and healing history. Infusions made from elderflowers are gentle and kind to the skin while having a mildly bleaching effect, and they can be used in the full range of skin preparations for face and body. As elder is both mild and healing, the infusion is frequently used to overcome skin problems especially when dealing with dry or sensitive skin. It may also help to reduce freckles.

Use rinses containing the flowers to lighten and condition graying and blonde hair.

elderflower face freshener

A gentle, softening, and antiseptic refresher suitable for balanced or slightly sunburned skins.

- 4 tablespoons elderflower infusion
- 4 tablespoons glycerin
- 1 tablespoon orange flower or rose water
- ½ tablespoon pure lemon juice

Combine all the ingredients together in a bottle and shake well. Seal and keep refrigerated.

elderflower lotion

Windburn is as painful and damaging as sunburn. This lotion can be used all over the body to ease both problems.

- 5 tablespoons elderflower infusion
- 5 tablespoons glycerin
- 3 tablespoons witch hazel
- 1 tablespoon almond oil
- 1 tablespoon eau de Cologne
- ½ teaspoon borax

Shake all the ingredients together in a bottle. Seal tightly and shake before use. Keep refrigerated.

Both marigold or chamomile infusions can be used as an alternative to elderflower.

superb elder conditioner

A smooth night cream suitable for dry and normal skins. Alternative infusions such as rose, chamomile, or lavender can be used to suit all skin types.

- 4 tablespoons emulsifying wax
- 2 tablespoons grated beeswax
- 3 tablespoons anhydrous lanolin
- 4 tablespoons each sunflower oil, almond oil, and sesame oil
- 2 tablespoons avocado oil
- 5 tablespoons elderflower infusion
- ½ teaspoon borax

Melt the waxes together in a *bain marie* and add the lanolin and oils. Warm the elderflower infusion in another bowl with the borax. Remove both bowls from the heat and slowly add the water to the oils beating constantly. Continue beating until the mixture is cold. Pot into sterile jars and seal.

elderflower hair rinse

This is a gentle rinse, with a delicate perfume, which will condition while adding soft highlights to fair hair.

- a good handful of dried elderflowers
- 1 cup distilled water

Simmer the flowers in the water for 30 minutes. Cover, and leave to cool and strain before using as a final rinse.

comfrey

Comfrey is a healing and antiseptic herb included in any skin preparation designed to counteract skin problems, specifically acne, blemished skin, eczema, rashes, and dry, flaking skin. Toners and cleansers using comfrey infusion are gentle enough to be used on delicate skins and powerful enough to act on problem oily ones. Comfrey infusion or decoction will also heal scalp problems when rubbed into the scalp or used as a final rinse.

comfrey face mask

A triple action mask which draws, cleanses and heals. It's best to avoid using this mask just prior to that special date.

- 1 teaspoon each of clear honey and brewer's yeast
- 1 teaspoon plain yogurt
- 1 teaspoon each of comfrey and marigold infusion
- 1 teaspoon olive oil

Thin the honey with a few drops of boiling water and blend in the yeast. Add the yogurt and the herb infusions, stirring well until it becomes a thick paste. Cover the face with a thin film of oil and then apply the mask. Leave for approximately 15 minutes to become dry. Wash off with tepid water, pat dry and apply a mild herbal toner.

comfrey and garlic poultice

This is one remedy for nasty pimples which works very well but entails the need for utter privacy because of the smell.

- 1 handful of fresh comfrey leaves
- 1 clove garlic
- Clear honey

Pound the leaves with enough warmed honey to make a good paste. Mash the garlic clove well and stir it into the mixture. Apply it using a warmed piece of soft dressing. Leave on overnight.

comfrey cleansing cream

A gentle, soothing cleansing cream. Elderflower, chamomile, chervil, and lady's mantle could all be substituted for comfrey.

- ¼ cup almond, sunflower, or olive oil
- ½ oz beeswax
- 2 tablespoons cocoa butter
- ¼ cup strong comfrey infusion
- 1 teaspoon borax
- 2 teaspoons clear honey
- 2 drops tincture of benzoin

Melt the oil, beeswax and cocoa butter together in a *bain marie*. Warm the infusion in another bowl and add the borax and honey stirring well until they dissolve. Remove both bowls from the heat and beat the comfrey mixture into the oils, adding the benzoin. Continue beating until the mixture thickens and cools. Pot in a sterile container and seal.

comfrey toner

A gentle toning lotion which is more appropriate for mature skins and those with an outbreak of blemishes. Adding the essential oil of rose geranium will also help smooth out fine lines.

- 6 tablespoons infusion made from comfrey and lady's mantle leaves
- 4 tablespoons glycerin
- 4 drops essential oil of rose geranium (optional)

Combine all the ingredients together in a bottle. Seal and shake well. Keep in the refrigerator.

thyme

This is one of the most powerful of the popular herbs which is used in astringent and cleansing infusions and as an antiseptic conditioner for the hair. Face packs and steamers using thyme are particularly good for cleansing greasy skin that may have blocked pores. Because thyme is so pungent it makes a strongly perfumed herbal oil which can be used to great effect in body rubs and massage oils where it not only smells very pleasant but also eases aches and pains, improves the quality of the skin, and deodorizes. The essential oil of thyme smells very strongly of the true herb. It is antiseptic and used dermatologically to clear skin conditions.

thyme and fig mask

A cooling, moisturizing mask to use on overheated skin. It will also soothe chilblains.

- 4 fresh ripe figs
- 1 tablespoon clear honey
- 1 teaspoon dried thyme

Roughly chop the figs and simmer them in a covered pan with the honey and enough water to cover the mixture. When they are softened mash the fig and honey well with the crushed thyme. When the mixture is smooth and pastelike in texture apply to clean skin and leave for 20 minutes.

spicy body toner

A fresh, tingling body lotion that will start the day well.

- 1 tablespoon each of fresh chopped thyme, rosemary, and mint
- 1 pinch each of fresh grated orange and lemon peel
- 1 pinch grated nutmeg
- 5 tablespoons orange flower or rose water
- 2 tablespoons alcohol

Blend all the ingredients in a glass jar, shake very well and leave to macerate on a warm window-sill for at least one week. Strain and bottle.

thyme soap balls

This recipe can be varied with any herbs or perfumes of choice. It is the ground rosemary that makes the soap slightly abrasive.

- 8 oz grated olive oil soap
- 1 cup thyme infusion
- 1 oz ground rosemary
- 8 tablespoons clear honey
- A few drops of essential oil of thyme or rosemary

Bring the infusion to a boil, reduce the heat and whisk in the soap flakes followed by the rosemary. Warm the honey slightly and add it to the soap mixture stirring well. Add the essential oil. Pour into molds and leave to harden. Egg boxes with waxed paper are a simple alternative to soap-making molds. Remember, soap takes quite a time to dry out.

thyme and lemon lotion

Used twice a day to rinse the face, this is an excellent toner for skin with acne.

- 2 sprigs of fresh thyme
- 2 cups of water
- Juice of ½ lemon

Boil the thyme in the water for two minutes. Cover and leave to infuse for five minutes. Strain and add the lemon juice.

Medicinal herbs

THIS CHAPTER FOCUSES ON 16 OF THE MAIN MEDICINAL PLANTS commonly used in clinical practice today. They have been selected for their diversity of application while being readily available from good health stores and pharmacies.

The section has been specifically written to include a brief review of each plant's historical place in traditional medicine as well as some of the most up-to-date scientific and medical findings relating to the herbs' biological actions and potential interactions with conventional drugs. It must be remembered that plants contain many drug-like components and need to be treated with respect and taken with care. It is hoped that this section will inform and educate the interested reader in their safe and established use, while being a reliable source of information to all. ■

garlic *Allium sativum*

OUR GREAT ALL-ROUNDER *Garlic has been cultivated for centuries not only for its medicinal properties but also for its unique ability to invigorate our taste buds and fortify our bodies. The more we study this special member of the lily family, the more we find out that garlic can help us in many different ways.*

If we look to garlic's history we see it has always been one of those herbs considered a cure-all. Dioscorides, the Greek physician, recommended garlic for snake bites, rabid dog bites, bloodshot eyes, baldness, eczema, herpes, leprosy, scurvy, toothache, and dropsy. The English herbalist Culpeper reports a similar plethora of uses: "It provokes urine, and women's courses, helps the biting of mad dogs and other venomous creatures, kills worms in children, cuts and voids tough phlegm, purges the head, helps the lethargy, is a good preservative against and a remedy for any plague..."

When we look at the latest research we find a pattern of use emerging.

■ **Helps reduce high cholesterol** By taking a garlic tablet equivalent to 4000 mcg of allicin (one of the powerful active agents in garlic) a drop of between 10 to12 percent can be achieved in total cholesterol level. While the lowering effect of

garlic is not massive, it has other benefits that may be even more effective in preventing heart disease: it elevates the levels of healthy cholesterol, known as HDL, by about 10 percent.

■ **Lowers blood pressure** Just how garlic lowers blood pressure is somewhat of a mystery. One mode of action may involve its sulfur-containing chemicals and could be related to its lipid (blood fat) lowering effect (described above) but the exact mechanism is not known. Whatever the action, you could experience a drop of up to 30 mmHg in systolic pressure (the top value in a blood pressure result) and about a 20 percent reduction in the diastolic result.

■ **Reduces the "stickiness" of the blood** One of the true killers of our modern society is not necessarily high blood pressure but the stickiness of the blood. Having a viscous sticky liquid circulating around the body can cause trouble should it decide to gel up

somewhere important. Once a clot forms, the tissue beyond dies due to a lack of oxygen. Your doctor may recommend a mini aspirin, while your naturopath will urge you to take extra garlic in the diet and a daily garlic pill to be on the safe side. Garlic is nature's dietary answer to conventional anticoagulant medication.

Modern clot-busting drugs can save the day in an acute emergency but for long-term management of clots, garlic may offer greater potential. Garlic and other natural agents such as the omega-3 fish oils, bromelain, and the herb *Capsicum* promote fibrinolysis (the dissolving away of a fibrin clot) and help prevent heart attacks and strokes.

■ **Enhances the immune system** Hippocrates described eating garlic to treat cancers. This was not such an odd suggestion when you consider how important the immune system is in controlling a cancerous condition. Human studies have clearly shown the great benefit garlic has on the ability of the immune system to fight infection and keep cancers at bay. The allicin component of garlic appears to have the most powerful anti-cancer effects.

- **Antibacterial activity** Garlic juice is known to kill many bacteria even the troublesome *Staphylococcus, Streptococcus* and *Brucella* strains. More recent studies have not only confirmed garlic's powerful antibiotic activity but have gone on to show that unlike conventional antibiotics the risk of the bacteria becoming resistant is negligible.

- **Antifungal actions** Unlike the powerful anti-fungal drugs, garlic is effective and very safe to use in any case of fungal infection. One of our common, unwelcome bugs is the yeast-like fungus called *Candida albicans*. Garlic has been shown to be more effective than the conventional therapy (nystatin), while building up the immune response to reduce the risk of re-infection.

- **Antiviral effects** There are very few effective anti-viral agents around so it is a blessing to have a few cloves of garlic to hand when influenza forces you to bed. The active agents in garlic, known as allicin, have strong viral-killing actions. In studies the herb has killed off the herpes simplex virus (types 1 and 2), parainfluenza virus, vesicular stomatitis virus and the rhinovirus responsible for the common cold.

- **Antihelminthic (worms) properties** It is true, just as our forefathers stated "... kills worms in children...," garlic has been shown on numerous cases to shift roundworm and hookworm infestations.

- **How to take garlic** Garlic can be taken in tablet, capsule or tincture forms. The odor on the breath is the deciding factor when choosing a garlic preparation. If you want to avoid a garlic odor, choose a strong standardized preparation that contains a high dose of allicin, in the region of 3.4 percent. Otherwise try the tincture form or any of the many proprietary garlic capsules.

- **Toxicity** There are very few adverse reactions to garlic. Large doses of garlic supplements can give stomach upsets or diarrhoea but this is simply cured by reducing the dose.

- **Drug-herb interactions** Very high doses of garlic may interfere with blood thinning drugs like Coumadin. One report showed that blood-clotting time doubled in patients taking a combination of garlic and Coumadin.

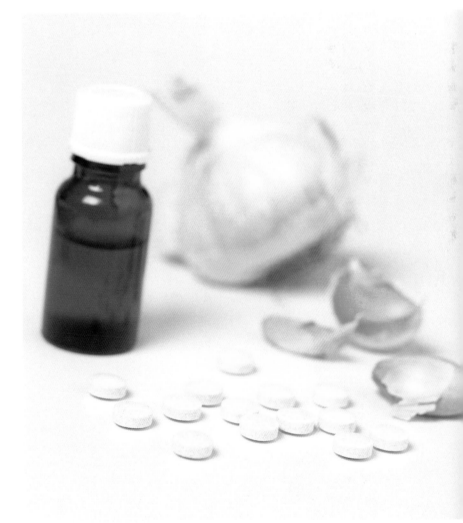

aloe vera *Alo barbadenis*

NATURE'S ALL-ROUND SOOTHER *Most common skin problems benefit from the soothing properties of aloe vera, and it is now known that aloe vera extracts offer a powerful stimulation to our immune system, going to the heart of our ability to fight infection.*

■ **Aids skin problems** The gel obtained from the aloe vera leaf appears to stimulate the healing process. Recent research has shown that specific fatty acids and hormone-like substances (called prosta-glandins) have been discovered in aloe vera extracts. These active agents along with naturally occurring vitamins C, E and zinc are now linked to its wound-healing properties.

Aloe vera has been used to successfully treat many skin problems, ranging from burns experienced by radiotherapy patients through to slowly healing leg ulcers and simple cases of eczema or dry patches of skin. However, it has been noticed that the gel can delay the healing of deep vertical cuts such as those associated with abdominal surgery, making this the only known situation where aloe gel should be avoided.

■ **Balances digestion** The internal use of aloe vera, in its juice form, can offer great relief to those with poor digestion or an ulcerated stomach.

By helping to regulate the stomach acid levels, a half-cup of juice taken for just one week can improve digestion and reduce the bloating often associated with bacterial or yeast overgrowth.

■ **Antidiabetic effects** Aloe vera has the ability to lower the blood sugar level. This was discovered during studies on the healing of leg ulcers in diabetic patients. It was noticed that blood sugar levels were affected by the aloe vera intake. This may help explain why diabetic patients have reported less of a need for their medication when using aloe vera products. However, one should be aware of the drug-herb interactions that may occur in diabetics.

■ **Fighting infections** At the heart of our ability to fight infection is the immune system which consists of specialized white blood cells. One of the key players in this white cell army is a cell known as the macrophage. It is now known that aloe vera extracts activate the macrophage cells and, in so doing, offer a powerful

stimulation to the immune system in general. Aloe vera, therefore, has the ability to activate the body's own killing machine, the immune system, while directly destroying many bacteria and viral invaders.

■ **How to take aloe vera** One of the original uses of aloe vera was simply to apply the pure gel to the skin for the treatment of wounds, local skin infections, dermatitis, and burns.

When taken internally it is best to obtain a commercially produced juice. This ensures the purity and correct dilution ratio. Always follow the instructions on the product or consult an experienced health professional.

■ **Toxicity** Some people with very sensitive skins may develop a rash when exposed to aloe vera gel. Applying a small patch test is always advisable before using the gel on a large area of skin.

■ **Drug-herb interactions** At a dose of 15ml (1 teaspoon) twice daily the gel or juice may increase the hypoglycaemic effects of anti-diabetic drugs such as glibenclamide.

angelica *Angelica sinensis*

THE GREAT HORMONE REGULATOR *Angelica is sometimes known as female ginseng, suggesting that the herb has its roots in traditional Chinese medicine. Indeed, angelica root became almost as popular as ginseng in traditional Chinese prescriptions.*

There are many different species of *Angelica* grown all over the world from China through to the USA and Europe. However, the most popular medicinal species are *Angelica sinensis* and *Angelica acutiloba*, both of which appear to have very similar actions and potencies and depend on the concentration of a plant chemical known as coumarin.

Helps hormonal imbalance The secret behind angelica's ability to effectively treat female hormonal problems lies in its high concentration of phytoestrogens. These plant-based substances are 400 times less active than the estrogens naturally found in the body, yet they exert a strong regulating effect on those tissues that normally respond to hormone estrogen. This helps explain how angelica can affect conditions that are characterized by both high and low estrogen levels.

In cases of menopause, characterized by low circulating estrogen, angelica can help stimulate the estrogen receptors on many tissues, including the reproductive tract. Alternatively, when estrogen levels are high or otherwise out of balance, as can occur in menstrual disorders, the phytoestrogens appear to work by blocking estrogen and its action on the tissues.

Research has now confirmed that the herb can help to relax smooth muscle which forms the wall of the uterus and other internal organs. When taken over the week before a menstrual period, angelica can help with abdominal cramps experienced by many women suffering from painful periods.

Immune regulator Traditional Chinese medicine has always formulated its prescriptions for cases of allergy to include angelica. Such ancient wisdom can now be supported by the finding that one of angelica's active agents, coumarin, enhances the activity of immune cells such as the macrophage. Such a powerful immune stimulation may also help explain how angelica-based medicines have been effective in destroying cancer cells and tumors. Another interesting aspect of its action on the cells of the immune system is an increased interferon secretion. Interferon is a special substance produced in the body that helps reduce the rate of cancer cell growth.

How to take angelica As with all herbal products, follow the manufacturer's instructions on remedies obtained from health stores or pharmacies. However, for a tincture form (normally in 65 percent alcohol) take 15 to 20 drops three times daily or one capsule containing 250 mg of the powdered extract twice a day.

Toxicity There have been no cases of acute toxicity reported for angelica although a light-sensitive rash may develop in a minority of those taking the herb.

Drug-herb interactions No reported cases.

hawthorn *Crataegus oxyacantha*

NATURE'S HEART TONIC *For as long as the records go back, hawthorn has been associated with the treatment of heart trouble and disorders of the circulation. It can ease attacks of angina, and improve the heart's health by enhancing the nutritional status of the heart muscle itself.*

Long before modern scientists discovered active heart chemicals in hawthorn, the ancient healers used a guiding principle in herbal medicine known as the doctrine of signatures, which related the look, shape, and physical characteristics of a plant to the body system. In the case of hawthorn, the red berries were likened to the heart and their color to the blood. It may sound unscientific but many of the healing herbs owe their indications to this process of classification.

■ **High blood pressure** This is a controversial issue in herbal medicine, but it would appear that the herb has no direct effect on blood pressure. However, blood pressure is a function of heart activity and the power of the heart beat, so hawthorn probably has a mild, lowering effect through its stabilizing action on heart function. Either way, for hawthorn to take effect on blood pressure, a good two to three weeks' continuous use is necessary.

■ **Eases a sick heart** Unlike many of the modern drugs that can ease acute attacks of angina, hawthorn actually improves the heart's health by increasing the blood flow through the coronary arteries, so that oxygen and nutrients are effectively delivered to the heart muscle. One of the commonest causes of heart disease is a narrowing of these feeder vessels, resulting in a starvation effect on the muscle cells. In time these cells can die due to an acute shortage of oxygen (a heart attack) or slowly lose their ability to function, causing heart failure.

Hawthorn is not the treatment of choice in acute cases of angina or heart attack but it can offer great support for those suffering long-term congestive heart failure. For many people hawthorn has enabled them to reduce their doses of glycoside drugs, such as digoxin, thanks to the enhancing actions on the heart muscle.

■ **Calms a fluttering heart** One of the latest indications for using hawthorn in heart disease is in the treatment of cardiac arrhythmia – heart flutters. Irregular heartbeats can be very disturbing. Most do not indicate serious underlying heart disease but it is important to obtain an ECG before treatment is started. So long as irregular beats do not represent a diseased heart, taking some extra minerals such as calcium and magnesium along with a dose of hawthorn could calm a fluttering heart safely and effectively.

■ **Gives strength to vascular plumbing** Our vascular system could be easily likened to a home central heating system, with the pipes representing the arteries and veins and the boiler pump, the heart! If the pipes become thickened and clogged by sediment the pressure increases in the heating system and the pump is placed under great pressure and in time will wear out prematurely. This is exactly what happens in the body.

Thickened blood vessels cause high blood pressure and this in turn places stress on the heart.

Hawthorn extracts can stabilize the very substance of blood vessels – collagen. The plant chemicals, known as flavonoids, contained in hawthorn have a powerful vitamin P-like activity. Bioflavonoids, also known as vitamin P, are present in all fruits and vegetables, they have the unique ability to increase the body's absorption of vitamin C. Hawthorn extracts, therefore, boost vitamin C levels and help deliver nutrition to the walls of the blood vessels.

Some of the other active agents contained in hawthorn extracts are known as proanthocyanidins. These are also found in grape seeds and have a protective effect against degenerative disease, especially in the circulatory system. In short, hawthorn can not only improve the strength of our blood vessels, it can also help prevent the common degenerative process known as atherosclerosis.

- **How to take hawthorn** In general 15 to 20 drops of the traditional 65 percent alcohol-based tinctures can be taken 2 to 3 times a day. For those needing a stronger remedy, tablets and capsules are the preferred form but it is recommended to seek professional advice in this instance.

  Another way to take hawthorn is in the phytosome form. This process makes the herb highly absorbable, mixing it with a fatty substance called phosphatidylcholine. A 100 mg dose of hawthorn phytosome (standardized) is probably the strongest hawthorn extract available.

- **Toxicity** There was concern that certain substances in hawthorn could cause cancer. These belonged to the proanthocyanidin group of organic compounds. However, it was later discovered that impurities in the test sample were responsible, not the proanthocyanidins which have subsequently been found to be protective against cancer.

- **Drug-herb interactions** Hawthorn may enhance the effects of cardiac glycosides such as digitalis and thus reduce the risk of side effects associated with high dose therapy.

  The active substances in hawthorn increase the coronary artery dilation effects caused by the drugs theophylline, caffeine, papaverine, sodium nitrate, adenosine, and adrenaline.

echinacea *Echinacea purpurea*

NATURE'S ANTIBIOTIC *One of the most popular uses of echinacea is in the treatment and prevention of the common cold and influenza. With its powerful anti-viral actions, echinacea will boost our natural defences over the winter months.*

It was the Native American Indians that discovered the healing potential of *Echinacea* and used it to treat all kinds of injuries and illness. Over the years *Echinacea* has grown in popularity as an immune-boosting herb and is probably one of the most commonly taken medicinal herbs in modern society.

■ **Immune enhancing properties**
Echinacea is known to contain many active immune-stimulating agents and essential oils and to date there have been well over 350 scientific studies looking into its chemistry. Most have confirmed that echinacea can help regenerate damaged tissue, reduce inflammation and stimulate the body's defence mechanisms by acting directly on the white blood cells. The herb has been used successfully to treat many viral infections such as influenza, herpes and stomatitis virus.

One of the most popular uses of echinacea is in the treatment and prevention of the common cold and influenza. This is best achieved by using preparations made from the fresh pressed juice made from the leaves, combined with extracts obtained from the roots, since this has powerful anti-viral actions.

■ **Controls infection** Contrary to popular belief echinacea does not have a powerful antibacterial action when applied directly to an infected area. It would appear that its ability to control infections comes from its immune-stimulating action, not via a direct effect on the bacteria itself. However, by taking a preventative dose over the winter, echinacea appears to boost our natural defence mechanisms. This does not mean that you will be immune from all ills but you may get over them much faster.

Many people benefit from taking echinacea for a variety of medical reasons. Some claim that their arthritis is eased while others see their recurrent and persistent cold sores clear for the first time, however the overwhelming majority find it a great help for chest infections, colds and influenza.

■ **How to take echinacea** The best form is a liquid extract preserved in 22 percent ethanol. Many brands are now standardized. For example, a product stating that it contains 2.4 percent of beta-1,2-fructo-furanoside, is desirable when selecting an echinacea product.

In general ¾ to 1 teaspoon (3 to 4 mls) of tincture or ¼–½ teaspoon (1 to 2 mls) of a fluid extract should be taken 3 times daily.

For children, give half this dose and for the very young (the under fives) consult a naturopath or herbalist.

Tablet and capsule versions are also available–follow the manu-facturers' instructions at all times.

■ **Toxicity** Echinacea has always been shown to be non-toxic at the recommended doses. There have been no reports of acute or long-term toxic reactions.

It is now generally accepted that echinacea works best when taken regularly for 8 weeks or so followed by 1 to 2 weeks off the remedy. Just like any therapeutic agent, natural or otherwise, the body can get used to the continued exposure and become tolerant to it. By giving the body a rest this problem can be avoided and the remedy will continue to work well.

■ **Caution** Even though echinacea is essentially a non-toxic herb, those suffering from auto-immune illness such as multiple sclerosis, Lupus, AIDS or other severe progressive illness should consult a health professional before taking this or any other herbal remedy.

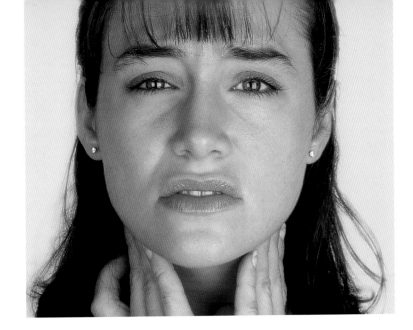

ginseng *Eleutherococcus senticosus*

THE STRESS BUSTER *Ginseng not only improves the general feeling of wellbeing and lifts the energy of those with chronic fatigue, it has a strong stimulatory effect to assist the body to resist the rigors of physical stress. Ginseng is an old answer to a modern health issue.*

Ginseng is a shrubby plant whose roots contain the biologically active agents. Most medicinal prescriptions cash in on the properties of the root but some may use the leaves. Interestingly, the roots contain the highest level of active agents just before the plant comes into flower. Analysis of the plant has shown that the principal drug-like agents belong to a large group of organic substances called eleutherosides. These give ginseng its famous stimulant and tonic effects.

Chronic fatigue syndrome For anyone suffering from chronic fatigue syndrome the main problem is simply finding enough energy to function from day to day.

Chronic fatigue has been around for many years but it has only recently become a recognizable medical condition, with the result that many more cases are now being identified.

One of the key issues in chronic fatigue is a lazy immune system. Most sufferers complain of recurrent health problems such as coughs and colds, chest infections, and sinus congestion. The relationship between the emotions and the state of the immune response is only now being fully appreciated and the use of ginseng's adaptogenic properties is becoming even more pertinent.

Adaptogenic properties have the ability to "normalize" a biological system. For example, when under physical or emotional stress the adrenal glands respond by releasing the hormone adrenalin. Ginseng can protect the adrenal glands and, despite continuous stress, help prevent stress-induced adrenal damage or exhaustion that may be in part responsible for chronic fatigue syndrome.

Ginseng has a strong stimulatory effect on the white blood cells known as T-helper cells. Once active these T-cells become highly effective against opportunistic infections.

Keeps stress in control In addition to countless anecdotal reports of the effect of ginseng in times of stress, there have been many interesting studies into this herb that show it to be very effective in improving tolerance to many stress factors, including noise and overwork.

How to take ginseng In general 15 to 20 drops of the traditional 33 percent alcohol-based tinctures can be taken 2 to 3 times a day.

For those needing a stronger remedy, tablets and capsules are the preferred form. For most users a standardized extract taken at a dose of 100 to 200 mg daily is adequate. Higher doses can also be taken but professional advice should be sought. A high dose regime should not be taken on a long-term basis because the body will become tolerant to its effects.

Another way to take ginseng is in the phytosome form. This process makes the herb highly absorbable by mixing it with a fatty substance called phosphatidylcholine. A 50 mg dose of ginseng phytosome is probably the strongest ginseng extract available.

Toxicity There have been no cases of acute or long-term toxicity reported from ginseng; however, some non-toxic side effects have been noted when taken at higher doses. These side effects include irritability and anxiety, headache and palpitations. It appears to be a case of less is more. Some patients with high blood pressure experience a lowering of the blood pressure when they take a lower dose over the long term, whereas their blood pressure may actually increase if they take a high dose.

Drug-herb interactions There have been a few reports of interactions with phenelzine, the monoamine xidase inhibitor.

OTHER COMMON PROBLEMS HELPED BY GINSENG

Angina	Cancer
High blood pressure	High cholesterol
Low blood pressure	Insomnia
Rheumatic disease	Hyperactivity in children
Chronic bronchitis	

ginkgo *Ginkgo biloba*

THE CIRCULATION BOOSTER *Gingko would appear to be one of the most popular medicinal herbs of all time. Its ability to improve short-term memory and clarity of thought, to aid sufferers from poor circulation, and to show promise in the treatment of Alzheimer's, puts ginkgo in steady demand.*

In 1988 German prescriptions of gingko topped 5.4 million. However, there have been concerns relating to the increase in demand since the ginkgo tree is slow growing and the harvesting of its green leaves has intensified greatly.

■ **Clears the head** One of ginkgo's names is the memory tree. It has long been noticed that those taking the herb report improvements in memory and a clarity of thought. Several scientific studies can now support this finding. Studies also indicate that the herb can be of help in cases of vertigo, tinnitus and some types of depression. Many age-related conditions have been shown to improve, especially those associated with impaired blood circulation.

One theory holds that as ginkgo increases the circulation to the brain, the levels of oxygen and glucose rise. These are two vital substances for optimal brain function and may be somewhat diminished in those suffering from poor peripheral circulation, such as the elderly or people with hardening of the arteries.

The treatment of Alzheimer's disease using ginkgo has been studied and some of the early results look promising, but reversing the effects of nerve destruction is beyond the scope of this herb.

■ **Helps cold hands and feet** Ginkgo appears to stimulate the circulation in the periphery, such as the hands and feet, as well as around the brain. Just why this is so is still unknown although studies have confirmed that the herb acts on the lining of blood vessels. Ginkgo has been shown to stimulate the release of special chemicals from the lining of the blood vessels. The end result is an opening up of blood vessels and an increased supply of oxygen and nutrition to the tissues of the body.

In one clinical trial, patients with poor circulation in the lower limbs were given 160 mg of a standardized ginkgo extract to take for two years. The results were astounding. Patients who could only manage to walk 200 feet could extend their pain-free exercise level to a 500-feet walk and their total walking distance to over 1,000 feet.

For those with less serious circulatory problems the extract may help ease the pain associated with Raynaud's disease, or simply help alleviate chilled hands and feet.

■ **Combats impotence** Known medically as erectile dysfunction, impotence is a distressing problem for many men. In most cases of erectile dysfunction the nervous system works well but the blood flow is impaired due to narrowing of the arteries. This may be age-related or due to underlying disease such as diabetes. Modern drugs like Viagra have revolutionized some men's sex life, but planning when to take the pill is not always convenient. Ginkgo is nature's Viagra due to its tonic effect on the erectile tissues and their blood supply. Taking ginkgo will not, however, yield instant results. After six months of taking 120 mg daily over 50 percent of those in a study reported improved erectile function. The longer ginkgo was taken the better the results.

■ **Failing eye sight** Another common age-related problem is macular degeneration. The macula is a special region of the light-sensitive lining of the eye, the retina. With advancing years the macula may suffer from a lack of blood supply and degenerate slowly, affecting the ability to read printed matter or see the fine detail of objects. Ginkgo has been shown to help slow this process and in some diabetic patients actually prevent it occurring in the first place.

■ **How to take ginkgo** In general 15 to 20 drops of the traditional 65 percent alcohol-based tinctures can be taken 2 to 3 times a day. For those needing a stronger remedy, tablets and capsules are the preferred form. Most of the research on ginkgo has focused on standardized extracts, which explains why some products state the percentage of active agents in their product. A good product will typically contain 24 percent ginkgoflavonglycosides at a dose of 40 mg. One tablet or capsule is normally taken 2 to 3 times daily. Higher 80 mg doses can also be taken but professional advice should be taken.

Another way to take ginkgo is in the phytosome form. This process makes the herb highly absorbable by mixing it with a fatty substance called phosphatidylcholine. An 80 mg dose of ginkgo phytosome (standardized to contain 24 percent ginkgoflavonglycosides) is probably the strongest ginkgo extract available.

■ **Toxicity** When taken in the recommended dose, ginkgo is a very safe herb. There have been over 44 scientific safety trials, involving some 10,000 people. Only minor side effects such as stomach upset, headache, and occasional dizziness have ever been reported.

It should be noted that the fruit of the ginkgo tree must be avoided. It is potentially toxic and even slight skin contact can produce a rash, itching, and blisters.

■ **Drug-herb interactions** Ginkgo may interact with anticoagulant drugs and enhance their activity. This effect has been noted in those long-term anticoagulant users who also took ginkgo for a long time. Another positive interaction of ginkgo is the improved effect it has on injected papaverine, a drug used to treat male impotence.

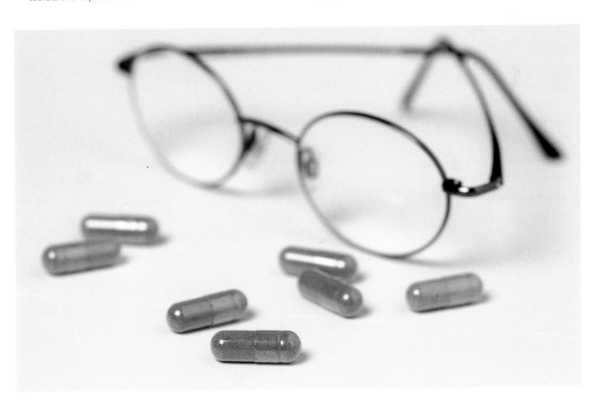

licorice *Glycyrrhiza glabra*

SOOTHES THE STOMACH AND GUT *For several thousand years licorice has been used in both western and eastern medicine to treat constipation, stomach problems, chest congestion and even malaria. Thanks to modern extraction methods, licorice-based remedies are now safe and effective for daily use.*

Interest in licorice started during the Second World War when a Dutch doctor treating stomach ulcers with licorice, saw many of his patients suffer from severe swelling of the face and limbs. The side effect was later identified to be due to a natural chemical in licorice, known as glycyrrhizin. Subsequent studies have shown that glycyrrhizin causes headache and fluid retention which can result in a rise in blood pressure. Modern licorice remedies have the glycyrrhizin removed, and the product is then known as degly-cyrrhizinated licorice (DCL) and carries no risk of side effects.

- **Heals an ailing gut** Licorice in its DGL form is a very effective healing agent for ulcers that occur in the stomach, intestines, and mouth. Compared to modern ulcer healing drugs, DGL stimulates the intestinal lining producing a protective barrier to digestive enzymes and acids. Under the influence of DGL the cells lining the gut multiply and grow over the ulcerated area. As they do so, licorice protects against the irritant effects of digestive juices.

  In other cases of gut upsets, such as gastritis (inflammation of the stomach), DGL can help reduce inflammation and protect the sensitive lining from further irritation from digestive acids.

- **Helps fight viral infections** There have been many medical indications for licorice. However, they rely on the glycyrrhizin content of the herb, which is unfortunately associated with adverse side effects. Some of these actions include antiviral activity against the herpes simplex virus and HIV infection.

- **May balance hormones** Other organic compounds in licorice such as isoflavones may have hormone-balancing effects that could help explain some of the reports relating to its traditional use in the treatment of menopausal and menstrual problems.

- **How to take licorice** Licorice in the form DGL is the preferred type. It is best to chew the DGL tablets very well since the activation of salivary enzymes and chemicals magnifies the healing powers of the licorice. The standard dose is 2 to 4 tablets of DGL (380 mg) chewed up and swallowed half an hour before food. This should be continued for at least 8 weeks but it may take up to 16 weeks for recovery to be complete.

  It is not recommended to chew or take the pure licorice root due to the high glycyrrhizin content.

- **Toxicity** Licorice in the DGL form is free from toxic side effects. High dose DGL may cause a loose stool.

- **Drug-herb interactions** Pure licorice may increase the loss of potassium when taken in conjunction with thiazide diuretics and increase the sensitivity to cardiac glycosides. This is due to the high glycyrrhizin content. However the use of DCL has practically eliminated any drug interactions.

St John's wort *Hypericum perforatum*

NATURE'S MOOD ELEVATOR *In many nations, sales of St John's wort extract far outstrip sales of regular anti-depressants, such that it has been dubbed "nature's Prozac." Some of the early uses of this herb as a balm relate to wound healing.*

Exactly how the name St John's wort came about may lie in the fact that the red oil which exudes from the buds and flowers is associated with the blood of St John the Baptist. Hippocrates, Dioscorides and Galen all make note of its wound-healing properties and its ability to help relieve nerve pains, such as sciatica.

Interest in St John's wort has risen recently since a number of medical trials have confirmed its ability to improve states of mild to moderate depression.

■ **Gives the mind a lift** In Germany sales of St John's wort outstrip sales of regular antidepressants, so much so that records show 66 million daily doses of St John's wort were prescribed in 1994. In over 26 clinical studies, all confirm the antidepressant effect. Nearly all of these studies have used a standardized extract (0.3 percent hypericin) at a dose of 300 mg taken three times a day.

■ **Antiherpes activity** Two of St John's wort's key compounds, hypericin and pseudohypericin, have attracted interest since they demonstrate powerful antiviral actions against herpes simplex and the HIV-1 virus.

■ **Heals and cleanses wounds** Historical records support the findings that infected wounds improve when St John's wort oil is applied. Analysis of the oil has shown that compounds known as phloroglucinols are responsible for this antimicrobial action. The same oil, when added to a cream base, can help accelerate the healing process in cases of minor wounds.

■ **How to take St John's wort** In general 15 to 20 drops of the traditional 65 percent alcohol-based tinctures can be taken 2 to 3 times a day. If, however you wish to take St John's wort for its antidepressant effect, the best preparation to use is 300 mg dose standardized to contain 0.3 percent hypericin extract.

It should be taken with food three times a day.

For topical applications massage in St John's wort oil 2 to 3 times daily.

■ **Toxicity** Acute toxicity is unlikely in the recommended dose, however a skin rash can occur on exposure to sunlight in some individuals.

Occasional stomach irritation has been reported when St John's wort is taken on an empty stomach. When taken at the full dose, cheese, wine, beer, pickled herrings, yeast, and yeast extract should be avoided.

■ **Drug-herb interactions** St John's wort can interfere with the metabolism of some drugs. Research has shown that the active principles affect the liver and a special chemical pathway involved in drug detoxification. Some of the drugs affected by this include antihistamines, oral contraceptives, and anti-epileptics. People taking selective serotonin re-uptake inhibitors (SSRIs) such as Prozac should avoid taking St John's wort.

peppermint *Mentha piperita var. vulgaris*

AIDS DIGESTION AND BEATS COLIC *Peppermint is so popular it has found its way into many aspects of our lives. It is an important medicinal, culinary and cosmetic herb. However, no record of peppermint is found in ancient medicinal texts simply because it was not discovered until the late seventeenth century*

Our ancient ancestors in general have used mints. Spearmint in particular has been mentioned in the Greek, Roman and Egyptian writings as a treatment for digestion.

- **Unwinds the spasm of irritable bowel syndrome** Irritable bowel syndrome (IBS) is one of the most common digestive problems in modern society. It lies in a disregulation in the nervous supply to the bowel, causing cramping, wind, and abdominal bloating. It is not that IBS sufferers make more bowel gas, it is rather that their bowel is overly sensitive to the normal peristaltic movements caused by gas production. The gut of an IBS sufferer reacts to these sensations by going into spasm, causing urgency, a loose stool and an apparent increased passage of bowel gas. Once the gas or stool is passed, the pain is relieved. To help reduce this distressing problem, pure peppermint oil is commonly used. The antispasmodic effects of peppermint oil appear to be due to

compounds known as polymerized polyphenols, but the other active principles such as menthol, menthone and methyl acetate probably play a vital role as well.

- **Combats gallstones** Gallstones may be susceptible to a combination of naturally occurring chemicals in peppermint oil. Peppermint can reduce bile cholesterol levels while increasing bile acid and lecithin levels in the gall bladder. Treatment can safely be continued for years.

- **How to take peppermint** As a simple tea, peppermint can soothe the stomach and reduce the incidence of indigestion (dyspepsia). Used as a muscle rub for sports injuries or around arthritic joints, peppermint (menthol) extract can help ease pain and stimulate the circulation.

  When treating the pain and spasm of colic or IBS-related pains, use an enteric-coated capsule. This guarantees the oil will reach the lower bowel to deliver the antispasmodic actions directly to the

problem area before the capsule opens. Normal capsules just give you minty breath! The enteric-coated capsules should contain 0.2 ml of peppermint oil. Take 1 to 2 capsules between meals twice daily.

- **Toxicity** Peppermint is generally considered safe. However, some individuals may be sensitive to the oil and develop a non-toxic rash, heartburn, occasionally a slowing of the heart rate, and very rarely muscular tremors. Topical applications can induce a hyper-sensitivity rash in some. If in doubt apply a small patch test before using peppermint or menthol extracts over a larger area.

- **Drug-herb interactions** People with achlorhydria (lacking digestive acid), and taking H2 receptor blockers e.g. Cimetidine or Ranitidine, should avoid peppermint tinctures or capsules. In such cases peppermint should only be taken in enteric-coated capsules.

kava kava *Piper methysticum*

A TONIC FOR THE MIND AND EMOTIONS *Kava kava is used to soothe the nerves, counteract fatigue and weight loss, and treat rheumatism. It has been tested against drugs such as valium in cases of anxiety and the results have been very impressive.*

Kava kava was a well-kept secret until the eighteenth century when the first Europeans made contact with Polynesian islanders who held this local plant in high esteem. Just as with alcohol in western societies, kava was associated with rules and customs of preparation known as the Kava Ceremony.

■ **Don't panic – reach for kava**
Kava is especially effective in reducing anxiety states and it has been used to treat insomnia and restlessness as well. Kava has been tested against drugs such as valium (diazepam) in cases of anxiety, and the results in all cases have been very impressive. In one study, 84 anxiety sufferers took the herb and reported improved memory, reaction times, and a feeling of "ease." Other studies have confirmed these findings with the additional benefit that kava proved non-habit forming and free of the complications associated with long-term diazepam use.

At the heart of kava's success lies a group of chemicals known as kavalactones. When kava extracts were standardized to contain 70 percent kavalactones, a significant benefit was observed in anxiety sufferers. This makes kava a really effective alternative for patients wanting to avoid benzodiazepine drugs, such as diazepam.

■ **How to take kava** In general 15 to 20 drops of the traditional 65 percent alcohol-based tinctures taken 2 to 3 times a day is adequate for mild anxiety or nervousness. For those needing a stronger remedy, tablets and capsules are the preferred form. Commercially prepared remedies should be standardized to contain 30 percent kavalactones. Doses of 200 mg are recommended 2 to 3 times daily. As with all herbal remedies taken at higher doses, it is recommended to seek professional advice.

■ **Toxicity** No side effects have been reported at the recommended dose range. Higher doses taken over a long time, however, may cause a skin rash that characteristically presents with a scaly dermatitis over the palms of the hands, soles of the feet, and the skin of the forearms, back, and shins. However, no cases of such skin reaction have been reported when taking a standardized extract.

■ **Drug-herb interactions** Kava may increase the effects of drugs that act on the central nervous system such as alcohol, barbiturates, and other mind-influencing drugs. It may also have an additive effect when taken in conjunction with tranquillizers such as benzodiazepine. Kava might reduce the effects of levodopa used to treat Parkinson's disease.

saw palmetto *Serenoa repens*

A MODERN REMEDY FOR ENLARGED PROSTATE

Traditionally, folk use revolved around saw palmetto's berries' ability to ease irritations to the delicate mucous membranes lining the urinary and reproductive organs. Some ancient texts also describe its aphrodisiac properties.

Saw palmetto is a small palm growing in the West Indies and along the Atlantic coastline. In recent times the berries have provided great benefit to men suffering from the age-related enlargement of the prostate. This process is known medically as benign prostatic hypertrophy (BPH). The power of saw palmetto has been a well-kept secret for many years, but modern research has now established it as a popular herbal treatment for prostate swelling.

■ **Slows prostate overgrowth** A hormone called dihydrotestosterone (DHT), which is produced within the cells of the prostate from testosterone, regulates prostate growth. Saw palmetto extracts have the ability to block this trans-formation of testosterone to DHT and in so doing reduce the overgrowth of the prostate gland.

In order to achieve this benefit it is essential that the extract is pure and potent enough. When looking for a suitable herbal supplement check the label for the standardization. A good saw palmetto product should contain 85 to 90 percent fatty acids and sterols. When taken at a dose of 160 mg twice daily, saw palmetto outperforms the standard medicinal prescription (Proscar) in all clinical studies. Not only is saw palmetto less expensive and safer that Proscar, it is actually more effective.

In studies, saw palmetto improved the flow of urine in BPH sufferers by 38 percent after three months of use compared to 16 percent in 12 months of Proscar use.

■ **How to take saw palmetto** To obtain the greatest benefit always take a standardized product containing 85 to 90 percent fatty acids and sterols. From clinical studies the optimal dose would appear to be 160 mg twice daily.

■ **Toxicity** Detailed studies have shown saw palmetto extracts to be non-toxic.

■ **Drug-herb interactions** None known.

WHAT IS THE PROSTATE ?

The prostate is a single doughnut-shaped gland about the size of a chestnut. It produces a nutrient fluid that keeps the sperm alive while helping prevent infection. A healthy prostate can contribute about 30 percent of the sperm volume. Beyond the age of 40 years the prostate can start to show signs of enlargement. Some men are affected more than others. If this enlargement starts to interfere with the flow of urine, the following symptoms are experienced: excessive night time urge to pass water, painful or difficult urination, decreased urine flow. Should you suffer from such symptoms, consult a health professional before taking saw palmetto extracts.

milk thistle *Silybum marianum*

THE LIVER TONIC *The descriptive name milk thistle originated from, and may have stimulated interest in, the use of the herb by nursing mothers to encourage milk flow. Many herbals have called the fruits of milk thistle "seeds." The fruits do look like seeds, being small and hard (technically called achenes) but they are the plant's fruits.*

The fruits have been traditionally used medicinally but the herb went out of vogue in the early twentieth century only to re-emerge when special liver protecting chemicals were discovered in its fruit extract. The group of organic compounds known as silymarin have subsequently been identified as being responsible for the variety of liver tonic effects.

- **Helps a toxic liver** The liver is one of the fundamental keys to human health. Everything we consume (food, drink, and drugs) has to pass through the liver before it goes into our general circulation. The liver is literally the body's chemical processing house. It produces all the enzymes needed to deactivate dangerous poisons and keep the body alive from day to day. Understandably, it can get overworked and the body can suffer from a toxic overload. This may not be life-threatening but can cause us to feel generally low and out of sorts. When experiencing a hangover from over-indulgence in food and drink, milk thistle's tonic effects are recommended.

- **Helps a toxic liver** When milk thistle extract was scientifically tested something amazing happened. A dose of one of nature's most potent liver toxins called phallotoxin (obtained from fungi) was injected, followed by a dose of milk thistle. Due to the unique ability of milk thistle to protect the chemistry of the liver from the poison, no damage occurred. The outcome encouraged its clinical use in the treatment of active hepatitis and cirrhosis with excellent results. Clinical trials using milk thistle with phosphatidylcholine, in patients with chronic viral hepatitis, demonstrated an improvement characterized by a reduction in liver enzymes used to monitor the level of disease activity. This supported the findings that this herb can target inflamed tissue and help prevent further damage.

- **How to take milk thistle** In general 15 to 20 drops of the traditional 65 percent alcohol-based tinctures can be taken 2 to 3 times a day.

  For those needing a stronger remedy, tablets and capsules are available but the preferred form is achieved by combining milk thistle with phosphatidylcholine. A 100 mg dose of milk thistle bound in this way (standardized to contain 80 percent silymarin) taken three times daily is probably the strongest way to take this herb.

- **Toxicity** Milk thistle is remarkably free of toxic reactions. Even at high doses it is well tolerated with occasional loose stool noted.

- **Drug-herb interactions** The concentrated extracts from the fruit protect the liver from the adverse effects of simultaneously administered drugs. Studies show that a dose of 400 mg can protect against the toxic effects of butyrophenones and phenothiazines. It was also shown to reduce the toxic effects of phenytoin.

feverfew *Tanacetum parthenium*

THE MIGRAINE RELIEVER *This flowering herb looks just like a large daisy. But looks can be deceptive. Feverfew contains a unique combination of active plant chemicals making it the herb of choice for migraine and arthritis sufferers. Interestingly, it re-emerged as a popular remedy only in the 1970s.*

However, as its name suggests this was not the original use of the plant. Feverfew found its place in the herbal medicine chest of old due to its ability to "dispel fevers," a use that has now gone out of fashion in favor of more effective anti-fever remedies.

■ **Stop migraine before it starts** The key to feverfew's success in migraine prevention lies in the substance known as parthenolide. This chemical has the unique ability to interfere with the "stickiness" of blood platelets (small cell fragments involved in the clotting function of blood), making them less likely to clump together. Feverfew also exerts a "tonic" effect on the smooth muscle that forms the walls of blood vessels. Additional actions include an anti-inflammatory mechanism. This has been linked to its ability to reduce the levels of circulating chemical irritants. All these substances are the focus of intensive research by major drug manu-facturers since they appear to hold the key to pain relief.

■ **Eases inflamed joints** In a similar way to feverfew's actions on migraine, inflamed joints also benefit from a dose of this herb. The same irritant substance responsible for the pain of migraine is also implicated in the swelling and pain associated with rheumatoid and other arthritic conditions. The use of strong anti-inflammatory drugs is often associated with significant side effects making them poorly tolerated over long periods of time. Feverfew may be an effective alternative.

■ **How to take feverfew** The effectiveness of feverfew depends on the concentration of the active principle parthenolide. An effective daily dose should be in the region of 0.25 to 0.5 mg of parthenolide. High quality feverfew products should display this standardization. A low daily dose may help to prevent migraine attacks but larger doses are needed during an attack. Experience will guide your own personal use but do not exceed 2 g in 24 hours.

■ **Toxicity** Chewing the fresh leaves may cause mouth ulcers and some people may develop dermatitis. In general the herb is well tolerated and without toxic side effects.

■ **Drug-herb interactions** No reported cases.

valerian *Valeriana officinalis*

NATURE'S ANSWER TO INSOMNIA *Insomniacs 1,000 years ago discovered the benefits of this plant and its herbal extracts. It was not until the mid 1980s that the active agents were thought to be the volatile oils. However, recent research has questioned the volatile oil theory throwing the origin of valerian's sedative principle into mystery once again.*

Most remedies are made from the dried root and rhizomes since the highest concentrations of calmative and tranquillizing principles are located in these tissues.

■ **Mild sedative to aid sleep** Sleep is a complex process and sleep disorders are very poorly understood. The use of a traditional herb such as valerian is a safe method in assisting the natural process without risking addiction or robbing the brain of its natural sleep-wake rhythms.

Studies carried out in sleep laboratories have confirmed that the herb exerts a mild sedative effect. The herb appears to act on the brain chemistry. GABA is a brain chemical responsible for relaxation and a feeling of ease. By stimulating the cells that normally respond to GABA, valerian can encourage rest and relaxation without forcing the body into a state of sedation.

It is important to appreciate that valerian is an effective herbal sedative but it is not strong enough to combat a high habitual caffeine consumption.

■ **How to take valerian** In general 15 to 20 drops of the traditional 65 percent alcohol-based tinctures can be taken 2 to 3 times a day. For those requiring a stronger remedy, valerian extracts standardized to contain 0.8 percent valeric acid should be taken as a dose of 150 to 300 mg. For best results take the herb 30 minutes before bed while avoiding caffeine and caffeine-containing food and drink, and alcohol.

■ **Toxicity** Valerian taken at the recommended dose is essentially a non-toxic herb.

■ **Drug-herb interactions** A recent review has found that valerian can increase the effect of barbiturates.

ginger *Zingiber officinale*

THE TRAVELER'S COMPANION *Like many herbs the medicinal use of ginger can be traced back to ancient China. Documented evidence dating back to 400 BC describes ginger being used to treat stomach, diarrhoea, nausea, rheumatism and toothache. Today it is the herb of choice for travel and morning sickness.*

The use of ginger in the treatment of morning sickness associated with pregnancy has become popular over recent years. A scientific study demonstrated that a 250 mg dose taken four times daily significantly reduced vomiting and nausea experienced by 19 out of 27 pregnant women studied. Because of the herb's exceptional safety record and extremely low level of toxicity it has entered conventional medicine's list of recommendations for morning sickness.

■ **Reduces nausea** For many years ginger has been recommended for motion sickness. NASA and the US Navy have both studied this herb with mixed results and reviews. The naval study found it helped its cadets find their "sea legs" quicker. It was noted, however, that one of the most important aspects of ginger related to the quality and purity of ginger powder used.

From the various studies it would appear that ginger acts on the stomach and gut rather than the nervous system. This may partly explain its positive effects on certain types of motion sickness and the variability relating to the type and dose of ginger preparation used.

■ **Beating the heat** Ginger has recently become a popular herb for the pain and inflammation associated with arthritis. It is now well established that chemicals belonging to the eicosanoid group cause the pain and inflammation of arthritis. These substances are active hormone-like compounds produced from dietary fat. They are eventually broken down in the body to even more potent inflammatory chemicals. Many of the commonly prescribed anti-inflammatory drugs block these chemicals and ease pain but their side effects, such as stomach irritation and ulceration, often limit their usefulness when it comes to long-term use. Ginger, on the other hand, has no adverse effects on the stomach and can effectively reduce the inflammatory effects of these compounds.

■ **How to take ginger** If taking ginger for travel or seasickness, take standardized 100 mg extract (look for products that contain around 20 percent pungent compounds), about four hours before traveling.

For morning sickness chop a slice of fresh gingerroot and crush it well to extract the juice. Mix to a paste with a freshly made cup of warm green tea. Sip slowly.

■ **Toxicity** When used in the prescribed dose as part of the general diet, ginger does not produce any known toxic effects. However, eating in excess of 6 g of ginger powder, some stomach irritation may be experienced.

■ **Drug-herb interactions** May enhance absorption of the drug called sulfaguanidine. Ginger may interfere with bleeding times. Doses in the region of 12 to 14 g may enhance the effects of anticoagulant drugs such as Coumadin.

Cooking with herbs

YOU DON'T HAVE TO BE AN HERB SPECIALIST TO KNOW THAT FOR the cook certain herbs are indispensable. They combine in exciting relationships and meld in perfect harmony: chives with new potatoes or eggs, dill with fish or cucumbers, tarragon with chicken, basil and tomatoes, rosemary and lamb. Parsley goes with just about everything and will enhance the flavor of other herbs in its company. Some are unusually high in nutrients. Using the recipes in this chapter, experiment and develop

 your own specialties. The only way to tell if the flavors are as you wish them is to taste as you cook.

The best way to experience herbs is freshly picked. Dried herbs lack the aroma of fresh, but are invaluable when fresh herbs are unavailable. Keep fresh herbs loosely wrapped in a plastic bag in the refrigerator for a day or so. Dried herbs should be stored in a dark, cool place in airtight bottles. Alternatively, green leaf herbs, such as chives or tarragon, can be frozen, simply by washing and drying them, then wrapping them in aluminum foil. They will stay flavorful for about two months. Use frozen herbs for cooking only, because while freezing does not ruin the flavor, it causes the leaves to go limp.

Dried herbs are more strongly flavored than fresh so a smaller quantity is required and can be added to a dish at the beginning of cooking, while fresh herbs are better mixed in toward the end of cooking. A *bouquet garni*, a term deriving from the French "garnished bouquet," is a bunch of herbs tied together and used in slow cooking to flavor soups and stews. Generally a bouquet garni uses robust herbs such as rosemary, thyme, bay, and parsley. To make, tie a sprig of parsley, thyme and a bay leaf together, and add a strip of lemon peel if a citrus flavor is required. ■

goat cheese tartina

Nothing could be simpler than fresh, tangy goat cheese spread onto bread and dribbled with olive oil infused with fresh herbs and garlic, and few things taste better.

SERVES 4

3 garlic cloves

Several pinches of salt

1 teaspoon lemon juice

2 tablespoons fresh, chopped chives

3 teaspoons each tarragon, parsley, dill, rosemary, and mint, as desired

⅓ cup olive oil

8 small, thin slices of rye bread, or French bread

Fresh, mild goat cheese, as desired

Crush the garlic with a mortar and pestle, and add the salt, then work in the lemon juice. Add the herbs and chives. Stir in the olive oil and leave to marinate for an hour or longer to allow the herbs' flavor to be absorbed.

Spread the bread with the goat cheese, then drizzle a little of the oil, either with the herbs intact or strained. Serve immediately.

scallops in saffron cream

Scallops are delicious with a mild creamy sauce which enhances but does not overpower their flavor. Steaming scallops ensures that they have a perfect texture and are not overcooked.

SERVES 4

20 fresh, prepared scallops

Juice and zest of 1 lemon

1 clove garlic, peeled and crushed

1 teaspoon fresh gingerroot, grated

For the sauce

¾ cup heavy cream

4 tablespoons dry white wine

4 tablespoons fish stock

1 tablespoon fresh chives, chopped

Few saffron strands

Rinse the scallops under running water and pat dry. Place in a shallow glass dish. Mix the lemon juice and zest, garlic, and ginger, and pour over the scallops. Stir to coat, cover and marinate for 1 hour, turning occasionally.

Remove the scallops from the marinade and transfer to a wax paper-lined steamer tier. Cover and steam for 3 to 4 minutes until the scallops are cooked through.

Meanwhile, heat the sauce ingredients in a small saucepan and cook for 5 to 7 minutes.

Spoon the scallops onto warmed serving plates and serve with the sauce, fresh steamed vegetables and rice or noodles.

sesame pikelets

Pikelets are a form of traditional pancake using a batter that is thicker than a pancake batter. This helps it to keep its shape while cooking. Make up the batter just before cooking as there is no need to let it stand.

SERVES 4

Pikelets

2 cups all-purpose flour

1 teaspoon sesame seeds

2 tablespoons butter, melted

⅔ cup milk

1 tablespoon light soy sauce

Topping

8 oz smoked trout fillets

2 tablespoons chopped fresh chives

1 tablespoon chopped fresh dill

⅔ cup sour cream

Lemon wedges, to serve

Sift the flour into a large mixing bowl. Add the sesame seeds and make a well in the center, gradually whisking in the butter, milk, and soy sauce.

Melt a little butter in a large heavy skillet. Drop 2 tablespoons of mixture into the skillet for each pikelet, cooking two at a time. Cook until the surface of the pikelet bubbles, then turn to brown the other side for 2 to 3 minutes. Cool on a wire rack. Repeat until all the mixture is used.

Slice the trout fillets. Mix half of the herbs into the sour cream. Spoon the sour cream onto the pikelets, top with the trout, and sprinkle on remaining herbs. Serve with lemon wedges, to squeeze over.

chicken dumplings with chives

Make preparation of these dumplings a family affair, and of course eating will be the best part.

SERVES 4

4 oz ground chicken

2 tablespoons light soy sauce

2 tablespoons ginger juice

¼ teaspoon white pepper

½ teaspoon dry sherry

1½ tablespoon cornstarch

2 cups chives, chopped into
   ½ inch lengths

1½ tablespoons sesame oil

24 dumpling wrappers

vegetable oil, for deep frying

Chili sauce, to serve

In a bowl combine the chicken, soy sauce, ginger juice, white pepper, and sherry, mixing them together well. Add the cornstarch and mix well, then set aside for at least an hour in the refrigerator.

In another bowl, mix the chopped chives and sesame oil, then set aside.

Just before starting to wrap the dumplings, mix together the chicken mixture and the chive mixture until thoroughly combined.

Place about one teaspoon of the filling onto a dumpling wrapper, brush the edges with a little water and pinch them closed. Repeat until the filling is used up.

Heat a wok or skillet and add oil to a depth of about 4 inches. Heat the oil to 375°F. Deep fry the dumplings, a few at a time, and turning them as they fry, for 3 to 4 minutes or until golden and crisp. Drain on paper towels.

Chicken dumplings can also be boiled for 5 to 6 minutes or steamed for 15 minutes.

Serve the dumplings with chili sauce.

Russian dilled potato salad

To reduce the calories and lower the cholesterol in this hearty dish, without diminishing the flavor, use non-fat or low-fat sour cream and yogurt. The addition of dill and chervil gives a subtle aniseed flavor.

SERVES 4

2 tablespoons white wine vinegar

2 tablespoons cider vinegar

1 tablespoon granulated brown sugar

1 teaspoon salt

1 teaspoon chervil, chopped

1 teaspoon wholegrain mustard

2 cups peeled and diced cucumber

1 cup plain yogurt

1 cup sour cream

1 tablespoon fresh lemon or lime juice

1 tablespoon dried dill

8 medium new, red potatoes

Salt

Mild sweet paprika

In a large non-metallic bowl, mix together the vinegars, brown sugar, salt, chervil, mustard, cucumber, yogurt, sour cream, lemon juice, and dill. Cover and chill in the refrigerator.

Wash the potatoes, but gently to keep the skins intact. Place the potatoes in a large saucepan, cover with water, and bring to a boil. Cook over a medium-high heat for 10 to 15 minutes, or until tender. Cool under cold running water and drain well.

Cut each potato into bite-size pieces. Fold the potatoes into the chilled yogurt and sour cream mixture and blend with a spoon.

Refrigerate for at least 6 hours to let the flavors blend. Season with salt and paprika to serve.

dill

salmon and broccoli penne

Steaming this delicious pasta sauce retains all the wonderful colors, textures, and flavors of the fish and vegetable. Use trout fillets in place of salmon if preferred.

SERVES 4

1½ cups dry penne

1 cup broccoli florets

½ tablespoon olive oil

1 teaspoon wine vinegar

2 garlic cloves, peeled and crushed

Juice and zest of 1 orange

12 oz salmon fillet, skinned and cut into cubes

6 tablespoons dry white wine

¾ cup light cream

2 tablespoons fresh dill, chopped

2 tablespoons freshly grated Parmesan

Salt and freshly ground black pepper

Orange wedges and dill, to garnish

Half fill the steamer base with water and bring to a boil. Add the pasta and season with a little salt.

Place the broccoli in a wax paper-lined steamer tier. Mix the oil, vinegar, garlic, and orange juice and rind together and pour over the broccoli. Rinse the salmon under running water and pat dry. Add the salmon to the pan and cover with a tight-fitting lid.

Set on top of the pasta in the salted boiling water for 10 minutes or until the pasta and fish are cooked through.

Meanwhile, heat the wine, cream, dill, and Parmesan in a pan without allowing it to come to a boil.

Drain the pasta and transfer to a warmed serving dish. Spoon the fish and broccoli on top and spoon the sauce over. Garnish, and serve.

smoked salmon and dill risotto

Smoked salmon is enhanced by the aniseed flavor of dill, and the feathery fronds make a glamorous garnish.

SERVES 4

3¼ cups fish stock

1¼ cups dry white wine

4 tablespoons butter

2 tablespoons lemon juice

1 red onion, cut into eight

2 garlic cloves, minced

2 cups arborio rice

Salt and freshly ground black pepper

1 teaspoon cayenne pepper

4 tablespoons chopped fresh dill

10 oz smoked salmon, cut into strips

⅔ cup light cream

Sprigs fresh dill, to garnish

Pour the stock and wine into a pan and bring to a boil. Reduce the heat to a gentle simmer.

Meanwhile, melt the butter in a large skillet and add the lemon juice. Gently fry the onion and garlic, stirring, until the onion has softened but not browned. Stir in the rice and cook gently, stirring, for 2 minutes until the rice is well-coated in butter.

Add a ladleful of the stock and wine mixture to the rice and cook gently, stirring, until absorbed. Continue adding the stock mixture until half of the stock has been used. Season well and add the cayenne pepper.

Continue adding stock for a further 20 minutes. Stir in the dill, salmon, and cream, and continue cooking, adding stock for a further 5 minutes until the risotto is thick but not sticky.

Serve in a warm bowl, garnished with dill.

baked lasagne with chervil

This dish actually tastes best with the strong, concentrated flavor of dried, rather than fresh, porcini, and the mushroom liquid can then be used in the sauce. While most recipes are deliciously adaptable to nearly any fungi, use only porcini in this one, for the finest fragrance and flavor.

SERVES 4 TO 6

3 oz dried porcini

2 cups water

8 shallots, chopped

5 tablespoons butter

3 tablespoons brandy

1½ cups light cream

Salt, ground black pepper, and a few
   gratings of nutmeg, to taste

10 oz pre-cooked lasagne

6 oz freshly grated Parmesan

5 tablespoons fresh chervil leaves,
   roughly chopped

Preheat a 375°F oven.

Place the porcini and water in a saucepan and bring to a boil. Reduce the heat and simmer over medium heat for 5 to 10 minutes, or until the mushrooms have softened. Remove the mushrooms from the liquid, chop roughly, and set aside. Strain the liquid for the sauce.

Sauté the shallots in the butter until softened, then add the mushrooms and cook for a few moments. Add the brandy, and cook over high heat until evaporated.

When the brandy has evaporated, ladle in ½ cup of the mushroom liquid, and cook over high heat until nearly evaporated. Repeat until all the liquid is used up and you have a concentrated, thin reduction. Now add the cream and simmer for 5 to 10 minutes. Season with salt, pepper, and nutmeg, and set aside.

In a 12 x 15 inch buttered baking pan place a layer of lasagne. Ladle in about a quarter of the mushroom sauce, a quarter of the cheese, and a sprinkling of the chervil. Repeat until mixture is used up, ending with the cheese and reserving the final sprinkling of chervil until serving.

Bake in the oven for 25 to 30 minutes, or until the cheese is melted and lightly browned in places. Serve immediately, with the reserved chervil scattered over.

roasted portobellos with pine nuts and chervil

*This delicious dish is easy to prepare. You may use large flat black
mushrooms instead of portobellos if you like. As well as chervil, use
fresh tarragon or chopped parsley and snipped chives.*

SERVES 4

4 portobello mushrooms or 12 large
 flat black mushrooms

5 garlic cloves, chopped

4 to 6 tablespoons olive oil, or to taste

2 tablespoons balsamic vinegar

Salt to taste

4 tablespoons pine kernels

1 to 2 teaspoon peppercorns, or to
 taste

1 tablespoon chopped chervil

Arrange the whole mushrooms in
a broiler pan or on a cookie sheet.
Sprinkle with the garlic, olive oil,
balsamic vinegar, and salt and
allow to marinate for 30 minutes.

Meanwhile, lightly toast the
pine kernels in a heavy-based, dry
skillet, over medium-high heat,
tossing every few moments, until
the pine kernels are golden and
lightly browned in spots. Remove
from the heat and set aside.

Broil the mushrooms in their
marinade, or bake in a 400°F oven,
until they are browned on their gill
sides. Then turn them over and
cook until the outsides are lightly
browned, about 10 to 15 minutes.

Arrange the mushrooms on
individual plates. Spoon over any
juices, scatter over the pine kernels,
peppercorns, and chervil and serve.

pan-fried chicken with tarragon

A simple but delicious supper dish when time is short. Serve with freshly cooked baby potatoes and sweet tender vegetables.

SERVES 4

4 boneless, skinless chicken breasts

A few sprigs fresh tarragon

1 tablespoon unsalted butter

1 cup white wine

2 shallots, peeled and finely chopped

²/₃ cup heavy cream

2 tablespoons chopped fresh tarragon

Salt and freshly ground black pepper

To serve

Freshly cooked baby potatoes and
   baby vegetables

Wipe the chicken breasts and make a few slits in each. Insert a few tarragon leaves in the slits and set aside.

Lightly brush or spray a skillet with oil then place on a moderate heat until hot. Add the butter and when melted add the chicken and cook for 5 to 6 minutes on each side or until done. Remove from the pan and set aside.

Add the wine and shallots to the pan and bring to a boil. Boil for 2 minutes, then reduce the heat and stir in the cream. Boil gently for 1 minute or until the sauce has thickened slightly.

Stir in the tarragon with seasoning to taste, heat for 1 minute, then serve with the cooked chicken and vegetables.

chicken and tarragon mushroom risotto

Tarragon is a classic flavoring for chicken. It is one of the subtlest of herbs and is a natural partner to chicken.

SERVES 4

5 cups chicken stock

1 stick butter

1 tablespoon oil

4 boneless chicken breasts, skinned

1 onion, finely chopped

2 garlic cloves, minced

4 large open cap mushrooms, peeled and sliced

2 cups arborio rice

1 tablespoon Dijon mustard

Salt and freshly ground black pepper

2 tablespoons chopped fresh or 1 tablespoon dried tarragon

4 tablespoons light cream

½ cup freshly grated Parmesan

Pour the stock into a pan and bring to a boil. Reduce the heat to a gentle simmer.

Meanwhile, melt the butter in a large skillet with the oil and cook the chicken for 5 minutes, turning until browned. Add the onion, garlic, and mushrooms and cook for 2 minutes until the onion has softened but not browned. Stir in the rice and cook gently, stirring, until the rice is well-coated in butter. Stir in the mustard.

Add a ladleful of stock to the rice and cook gently, stirring, until absorbed. Continue adding small quantities of stock to the rice until half of the stock is used and the rice is creamy. Season and add the tarragon.

Continue adding the stock until the risotto is thick but not sticky, about 25 minutes. Stir in the cream and cheese and serve in a warm bowl.

skate with tarragon butter

Skate is a wonderful-looking and tasting fish, but if unavailable, use meaty cod or monkfish in its place. Tarragon and fish is another perfect combination, but dill or fennel tops may be used instead for an equally good sauce.

SERVES 4

2 medium-size skate wings, about 2 lb

¾ cup dry vermouth

Juice and zest of 1 orange

1 leek, sliced

1 red onion, quartered and thinly sliced

For the tarragon butter

1 stick butter

2 tablespoons fresh tarragon, chopped

Few drops of Tabasco sauce

2 teaspoons mixed peppercorns, partly crushed

Salt

Tarragon sprigs, to garnish

Rinse the skate under cold water and pat dry with paper towels. Cut the skate wings in half to give four equal portions.

Line two steamer tiers with wax paper and place two portions of fish in each.

Mix the vermouth, half the orange juice and zest and pour over the fish. Add half the leek and onion to each tier and stack on top of each other in the steamer base. Cover with a tight-fitting lid and cook for 20 minutes, swapping the tiers over halfway through

cooking, until the fish is cooked through.

Meanwhile, melt the butter in a small saucepan and add the remaining orange juice and zest. Stir in the remaining sauce ingredients and heat gently for 2 to 3 minutes.

Remove the skate from the steamer and transfer to warmed serving plates. Spoon the butter over the skate, garnish with orange slices and fresh tarragon sprigs, and serve with freshly steamed vegetables including new potatoes.

hot and sour soup

Try eating this soup when you are feeling slightly under the weather or have the beginnings of a cold. The chiles in the soup help clear the head and make you feel better.

SERVES 6

¼ cup dried mushrooms

1 small chicken breast, skinned

6 oz tofu (bean curd), drained

3 cups chicken broth, preferably homemade

1 to 2 bird's eye chiles, deseeded and chopped

3 lemongrass stalks, bruised, outer leaves discarded

1 carrot, peeled and cut into thin strips

2 celery stalks, trimmed and cut into thin strips

3 tablespoons dark soy sauce

¼ cup snow peas, halved

½ cup bean sprouts

4 tablespoons cornstarch

2 tablespoons dry sherry

2 tablespoons chopped fresh cilantro

Soak the mushrooms in ⅔ cup almost-boiling water for 20 minutes. Drain, setting aside the mushrooms and the liquid. Chop the rehydrated mushrooms into small pieces if necessary.

Cut the chicken into thin strips and the tofu into small dice, and set them aside.

Heat the wok, then add the broth with the chiles and lemongrass, and simmer for 3 minutes. Add the mushrooms, liquid, chicken strips, tofu, carrot, celery, and soy sauce. Bring to a boil, and simmer for 2 minutes.

Skim if necessary and add the snow peas and bean sprouts. Cook for another minute.

Blend the cornstarch with the sherry, stir into the wok, and cook, stirring until slightly thickened.

Stir in the cilantro, heat for 30 seconds, and serve.

tomato and cilantro soup

This is a refreshing cold summer soup that would make an excellent first course to a fish or poultry main dish. The citrus-like flavor of the cilantro perfectly complements the fruit juices in the soup.

SERVES 6

3 lb ripe, plump tomatoes, roughly chopped

1 small onion, chopped

¾ cup tomato juice

3 tablespoons freshly squeezed orange juice

1 red bell pepper, deseeded

¾ teaspoon superfine sugar

Iced water

4 tablespoons chopped fresh cilantro

¾ cup plain yogurt

In a blender purée the tomatoes, onions, tomato juice, orange juice, bell pepper, and sugar.

Press the purée through a strainer, rubbing with a wooden spoon to force as much through as possible. Discard the residue, and add sufficient iced water to thin the purée to a soup-like consistency.

Stir in the cilantro, cover, and chill. Pass the yogurt at the table, to allow guests to add as they wish.

pan-fried salmon with black-bean relish

This warm, spicy relish with its rich partnership of flavors is the perfect complement to lightly fried salmon. For a milder relish, remove the seeds and veins of the jalapeños or omit them entirely.

SERVES 4

4 salmon steaks, about 6 oz each

Salt and pepper

1 tablespoon olive oil

For the relish

1 cup fresh or frozen corn kernels

15-oz can black beans

2 tablespoons olive oil

2 tablespoons lime juice

¼ cup diced red bell pepper

¼ cup diced green bell pepper

2 tomatoes, seeded and chopped

¼ cup chopped scallions

2 tablespoons chopped cilantro

1 or 2 jalapeño chiles, minced

¼ teaspoon ground cumin

¼ teaspoon salt

Pinch black pepper

Prepare all the relish ingredients.

Season the salmon on both sides with salt and pepper. Heat the olive oil in a large skillet. Add the salmon and cook over medium heat, turning once, until the fish just starts to flake and just a bit of dark pink flesh can be seen in the center, 3 to 5 minutes each side, depending on the thickness of the steaks.

While the salmon is cooking, heat the corn, black beans, olive oil, and lime juice in a pan until the corn is tender, about 5 minutes. Add the remaining ingredients and cook over medium heat just until warmed through.

Serve the salmon steaks topped with a large spoonful of relish.

steamed mussels with creamy fennel sauce

Fennel is a great accompaniment to succulent mussels, adding a subtle aniseed flavor that is perfect with seafood. Use white grape juice in place of wine as a non-alcoholic variation.

SERVES 4

4 lb mussels, scrubbed, beards
  removed

1 carrot, cut into julienne strips

1 fennel bulb

2 garlic cloves, minced

2½ cups fish stock

For the sauce

2 tablespoons butter

1 small red onion, finely diced

1 clove garlic, minced

¼ cup dry white wine

1 tablespoon fresh mixed herbs,
  chopped

Freshly ground black pepper

¼ cup heavy cream

Clean the mussels, removing the beards, and discarding any shells that do not shut tight when tapped. Place the mussels in the base of a steamer with the carrot strips.

Finely shred half the fennel and add to the pan with the garlic and stock. Cover with a tight-fitting lid and steam for 5 minutes.

Remove all the mussels and vegetables from the stock with a slotted spoon. Discard any unopened mussels. Keep warm. Measure up ¾ cup of the stock and set aside.

Meanwhile prepare the sauce. Melt the butter in a saucepan and sauté the onion and garlic for 2 minutes until softened. Add the wine, herbs, black pepper, and cream. Pour in the ¾ cup of fish stock.

Finely chop the remaining fennel and add to the pan. Bring to a boil and cook for 8 to 10 minutes.

Spoon the mussels into warmed serving dishes and top with the vegetables. Serve with the wine and cream sauce.

eggplant, fennel, and walnut salad

The salted walnuts really bring this salad to life, and complement the eggplant and fennel well.

SERVES 6

¾ cup olive oil

1 fennel bulb, finely sliced, feather leaves reserved for garnish

1 small red onion, sliced

¾ cup walnut pieces

Salt and freshly ground black pepper

1 large eggplant, cut into ½-inch pieces

1 tablespoon red wine vinegar

1 tomato, skinned, seeded and chopped

1 tablespoon torn fresh basil leaves

Heat 3 tablespoons of olive oil in a skillet and add the fennel and onion. Cook until just soft but not browned, about 5 to 8 minutes. Remove with a slotted spoon and place in a salad bowl.

Add 2 tablespoons of oil to the skillet, then stir in the walnut pieces and fry them for about 2 minutes, until crisp and browned but not burnt. Remove the nuts from the skillet with a slotted spoon and drain on paper towels. Sprinkle with salt and toss the nuts until well coated and cool.

Add 4 tablespoons of oil to the skillet, then add the eggplant and fry over a moderate heat until tender and browned on all sides. Remove from the skillet and add to the fennel and onion.

Add the remaining oil to the skillet with the red wine vinegar and a little salt and pepper. Heat, stirring, until it is simmering, then pour over the vegetables in the bowl. Toss lightly then leave to cool.

When the salad is still slightly warm, add the salted walnuts, chopped tomato and basil. Leave until cold, then serve garnished with fennel leaves.

scallop and fennel risotto

Scallops flavored with fennel and celery give this risotto a distinctly savory taste.

SERVES 4

2 tablespoons olive oil

1 lb scallops, halved or quartered if large

1 teaspoon celery seeds

5 cups fish stock

2 tablespoons butter

2 leeks, finely chopped

2 sticks celery, trimmed and finely chopped

1 bulb fennel, trimmed, finely sliced, and fronds reserved

2 cups arborio rice

Celery salt and freshly ground black pepper

½ cup freshly grated Parmesan

Chopped celery leaves, to garnish

Heat the olive oil in a large pan and gently fry the scallops and celery seeds for 3 to 4 minutes until done. Drain, reserving the pan juices, and keep warm.

Pour the stock into a saucepan and bring to a boil. Reduce the heat to a gentle simmer.

Melt the butter with the reserved pan juices in a large pan. Gently fry the leeks, celery, and fennel for 3 to 4 minutes until just softened. Add the rice and cook, stirring, for 2 minutes until well mixed.

Add a ladleful of stock and cook gently, stirring, until absorbed. Continue ladling in the stock until all the liquid has been absorbed and the rice is thick, creamy, and tender. Keep the heat moderate. This will take about 25 minutes.

Mix in the scallops and season. Heat through for 2 minutes. Just before serving, stir in the Parmesan. Serve garnished with chopped celery leaves and reserved fennel fronds.

charred cod with pesto

Basil leaves are the base ingredient of the Italian sauce, pesto. Enhanced here with pine kernels, it is an exciting accompaniment to pan-fried cod.

Serves 4

FOR THE PESTO

1 cup fresh basil leaves

2 tablespoons toasted pine kernels

4 garlic cloves, peeled and chopped

1 tablespoon lemon juice

½ cup olive oil

2 tablespoons freshly grated Parmesan

For the fish

Four 5-oz cod fillets or steaks

2 tablespoons olive oil

4 tablespoons lemon juice

3 garlic cloves, peeled and sliced

1 tablespoon roughly torn basil leaves

Place the basil leaves with the pine kernels, garlic, and lemon juice in a food processor and blend for 30 seconds. Keeping the motor running, slowly add the olive oil and then stir in the Parmesan. Scrape into a small bowl, cover, and set aside.

Lightly rinse the fish, pat dry, and place in a shallow dish. Blend the olive oil and lemon juice with the garlic and torn basil leaves and pour over the fish. Cover lightly and refrigerate for 30 minutes.

Lightly spray or brush a grill pan with oil then place on a moderate heat until hot. Drain the fish and cook in the pan for 3 to 5 minutes on each side, depending on thickness, until done.

Remove from the pan and spoon over a little of the pesto and serve with freshly cooked noodles, sliced tomatoes, and black olives. Garnish with basil leaves and serve the remaining pesto separately.

ratatouille and goat cheese quiche

A light, and simply delicious flan that would be perfect for lunch or as an appetizer for a vegetarian dinner.

SERVES 4 TO 6

Pastry
¾ stick butter
¾ cup fine whole wheat flour
pinch of salt

For the filling
14-oz can ratatouille
1 tablespoon basil leaves, roughly torn
⅔ cup goat cheese
salt and freshly ground black pepper
1¼ cups milk, or milk and light cream, mixed
2 large eggs, beaten

Preheat the oven to 400°F. Prepare the pastry by blending the butter into the flour and salt until the mixture resembles fine bread crumbs. Mix to firm, manageable dough with warm water, then knead lightly on a floured surface and roll out to line a deep 7-inch pie pan. Chill the pastry for 10 to 15 minutes, then line the pastry case with waxed paper and fill with baking beans. Bake for 15 minutes in the preheated oven.

Remove the paper and beans and reduce the oven to 375°F. Spread the ratatouille over the partly-cooked pastry. Sprinkle the basil leaves and goat cheese over and season with pepper. Beat together the milk, the eggs, and some seasoning, then pour the mixture into the pastry case over the vegetables and cheese.

Return the quiche to the oven, and cook for a further 35 minutes. The quiche is best served warm.

eggplant pesto

As an alternative accompaniment, this relatively low-fat version of pesto has a hint of smokiness from the eggplant.

SERVES 4

1 large eggplant

1 large handful of fresh basil leaves

3 garlic cloves, roughly chopped

½ cup pine kernels

¾ cup freshly grated Parmesan cheese

1 teaspoon coarse sea salt

¼ cup olive oil

Broil the eggplant until the skin is wrinkled and blistered and the flesh is tender, about 15 to 20 minutes. Cover with a damp cloth and leave to cool slightly for about 10 minutes, then peel off the skin.

Blend all the remaining ingredients together in a blender or food processor, then add the eggplant and blend again. Season to taste. Serve tossed into freshly cooked pasta.

tomato and basil bruschetta

A tasty bread to serve with fish dishes or salads for a fresh and interesting alternative to garlic bread, but just as garlicky.

SERVES 4

4 thick slices of Italian or French country bread

½ cup olive oil

6 very ripe, flavorful tomatoes, diced

Handful of fresh, sweet basil leaves, torn

4 garlic cloves, minced

Coarse sea salt, to taste

Brush the bread with several tablespoons of the olive oil, then toast it on a baking sheet at 425°F for about 15 minutes, turning once or twice or until crisp and golden brown.

Combine the tomatoes with the rest of the olive oil, basil, and garlic, sprinkle with coarse sea salt, and serve on toasted bread.

risotto alla bolognese

Usually associated with spaghetti, the rich sauce flavoring this risotto contains the classic bolognese ingredient, oregano.

SERVES 4

3¾ cups beef stock

⅔ cup red wine

1 stick butter

1 cup ground beef or veal

2 rindless bacon slices, chopped

1 onion, finely chopped

2 garlic cloves, minced

2 cups arborio rice

Salt and freshly ground black pepper

2 tablespoons tomato paste

7-oz can chopped tomatoes

1 carrot, diced

1 stick celery, sliced

2 tablespoons chopped fresh oregano

Pour the stock and wine into a pan and bring to a boil. Reduce the heat to a gentle simmer.

Meanwhile, melt the butter in a large skillet and gently cook the ground beef and bacon for 2 to 3 minutes until the beef is sealed. Add the onion and garlic and cook for a further 2 minutes, stirring, until the onion has softened but not browned. Stir in the rice and cook for 2 minutes, stirring, until the rice is well coated in butter.

Add a ladleful of stock and wine and cook gently, stirring, until the liquid has been absorbed. Continue adding stock until half of the stock has been used and the rice is creamy. Season well and stir in the tomato paste, tomatoes, carrot, and celery. Continue adding the stock until the risotto becomes thick but not sticky, about 25 minutes. Stir in the oregano to serve.

San Francisco cioppino

If San Francisco has a signature dish, it is cioppino, a wonderful seafood stew. Its origins are in a modest Italian fish soup, which in turn is a more rustic version of bouillabaisse. The broth is rich tomato, which usually includes wine and herbs. Be sure to provide tools for cracking crab shells, and finger bowls and plenty of napkins.

SERVES 10 TO 12

½ cup plus 2 tablespoons olive oil

1 large onion, chopped

3 leeks, white part only, chopped

1 red and 1 green bell pepper, chopped

8 garlic cloves, minced

1½ lb fresh tomatoes, peeled, deseeded, and chopped, or three 15-oz cans whole tomatoes, chopped

4 tablespoons chopped parsley

2 teaspoons dried basil

1 teaspoon dried oregano

½ teaspoon dried thyme

2 bay leaves

¼ teaspoon dried red pepper flakes, or more to taste

2 cups dry red wine

9 cups fish broth

Salt to taste

2 large crabs, cooked and cracked

2 dozen clams in their shells, scrubbed

2 dozen mussels in their shells, scrubbed

1 lb sea bass, swordfish, or other sturdy, non-oily fish, cut into 1-in cubes

1½ lb shrimp, shelled and deveined

In a very large pan or stockpot, heat ½ cup olive oil. Add the onion, leeks and peppers. Sauté for 10 minutes.

Meanwhile, heat the remaining 2 tablespoons olive oil in a small sauté pan. Add the garlic and sauté for 2 minutes. If the garlic browns too quickly, remove the pan from the heat. The residual heat in the oil will continue to cook the garlic. Add the garlic and oil to the stockpot.

Add the tomatoes, herbs, pepper flakes, wine, and fish broth to the pan. Bring the soup to a boil, then lower the heat and allow to simmer, uncovered, for 45 minutes. Taste, adding more red pepper flakes if a spicier broth is desired, and add salt if needed. If you are making the broth in advance, remove it from the heat, cool, and refrigerate. About 30 minutes before serving, reheat, bringing it to a boil.

Add the seafood to the broth: first, the crab, about 15 minutes before serving time. Add the mussels or clams about 5 minutes later. Add the fish 7 to 8 minutes before serving time, and finally, about 3 minutes before serving, add the shrimp.

Remove and discard the bay leaves, as well as any mussels or clams whose shells have failed to open. Put some seafood in each bowl, making sure everyone gets a selection, then ladle over the broth.

smoked haddock pots

These individual pots may be made with any smoked fish, such as cod or salmon for variation. Creamed cottage cheese is used in the recipe, and the fresh taste of parsley enhances the smoked fish.

SERVES 4

2 cups fresh spinach, washed and stems removed

8 oz smoked haddock fillets, skinned and flaked

1 egg

1 egg yolk

¾ cup creamed cottage cheese

½ cup plain yogurt

2 tablespoons fresh parsley, chopped

Juice and zest of 1 lime

Freshly ground black pepper

1 large tomato, thinly sliced

Lime slices and tomato, to garnish

Prepare the steamer, bringing the water to a boil. Steam the spinach in the top of the steamer for 2 minutes. Remove, and squeeze out any moisture.

Rinse the fish under running water and pat dry. Place with the egg, egg yolk, cheese, yogurt, parsley, lime juice and zest and pepper in a food processor and blend for 30 seconds.

Arrange half of the spinach in the base of four lightly buttered ramekin dishes and top with half of the fish mixture.

Layer the tomatoes on top, and then the remaining fish mixture. Top with the remaining spinach.

Place the dishes in the top of the steamer, cover with a tight-fitting lid and steam for 20 minutes until set. Invert the ramekins onto serving plates, garnish and serve with hot or toasted bread.

fattoush

This popular Lebanese salsa has pieces of crisply toasted pita bread added just before serving. This allows the pita to soak up some liquid without becoming soggy.

SERVES 4 TO 6

1 cucumber, diced

1 large red bell pepper, cored, deseeded, and diced

4 ripe tomatoes, diced

½ cup black olives

Bunch of scallions, thickly sliced on the diagonal

2 tablespoons chopped fresh flat-leaf parsley

2 pita breads, toasted until crisp and golden

Juice of ½ lemon

3 tablespoons olive oil

Salt and freshly ground black pepper

Toss together the cucumber, pepper, tomatoes, olives, scallions, and parsley in a large bowl. Tear the pitas into bite-sized pieces and add to the cucumber mixture.

Whisk together the lemon juice, olive oil, and plenty of seasoning. Pour over the salad, toss well together and serve immediately.

pasta primavera

A celebration of the first months of summer when asparagus, peas and beans are small, bright green and full of flavor

SERVES 4

1 cup sugar snaps or snow peas, topped and tailed

1 cup fine green beans, topped, tailed and halved

1½ cups asparagus, trimmed and cut into 2-in lengths

1 cup shelled fava beans

1 small leek, sliced finely

1 tablespoon butter

1 cup heavy cream

8 oz whole wheat pasta

1 to 2 tablespoons freshly chopped parsley

Bring a pan of salted water to a boil. Cook the sugar snaps, beans, and asparagus individually, plunging them into iced water immediately to prevent over-cooking. Cook the asparagus for 3 minutes, then add the tips and cook for a further 2 minutes; cook the fava beans for 3 minutes, sugar snaps for 2 minutes, and the fine green beans for 1 minute.

Cook the leek slowly in the butter until soft but not brown, then add the cream and heat until almost boiling. Drain the vegetables and add them to the pan then heat gently for 2 to 3 minutes until piping hot. Stir in the parsley.

Cook the pasta in a large pan of boiling salted water, then drain and shake dry. Add the pasta to the vegetables, tossing it in the cream, and serve immediately.

fried parsley and celery leaves

The way the flavor is concentrated when these leaves are crisply deep-fried is amazing. Don't allow them to overbrown.

Fry a test piece of parsley or celery leaf in the oil to check the temperature. If it is too hot, the leaves will burn before you can remove them and if it is too cool they will remain soggy.

Remove the stalks from a large bunch of dry, clean parsley and some celery tops. When the temperature is right, drop the leaves into the oil until the crackling noise stops and then remove quickly with a slotted spoon. Drain on paper towels.

lamb steaks with garlic and rosemary

This recipe is very quick and easy to prepare and cook. If time is really short you can cut down the marinating time by half.

SERVES 4

Four 4-oz boneless lamb steaks

3 garlic cloves, peeled and cut into slivers

A few small fresh rosemary sprigs

1 tablespoon grated orange zest

2 shallots, peeled and sliced into wedges

2 tablespoons olive oil

3 tablespoons red wine vinegar

Rosemary sprigs and orange zest, to garnish

Wipe the steaks and, using a sharp knife, make small slits on both sides then insert the slivers of garlic and rosemary. Place in a shallow dish and sprinkle over the orange zest and shallots. Blend the oil and vinegar and pour over the steaks. Cover lightly and leave in the refrigerator for 30 minutes, turning the steaks occasionally.

Lightly brush or spray a grill pan with oil then place on a moderate heat until hot. Drain the steaks and cook in the pan for 4 to 6 minutes or until done to personal preference. Garnish with rosemary sprigs and orange zest and serve with sauteed potatoes and ratatouille.

rosemary-roasted veal

Braised mushrooms added to the pan juices make a luscious sauce for roasted meat, flavored with rosemary.

SERVES 6

2½ lb lean veal, rolled and tied

Several sprigs fresh rosemary

10 garlic cloves, half cut into slivers, half chopped

Salt and ground black pepper

4 tablespoons olive oil

2 carrots, diced

10 garlic cloves, left whole but peeled

1 large onion, chopped

3 fresh ripe tomatoes, finely chopped

1 lb mixed fresh mushrooms

Dry white wine or stock, if needed

Preheat oven to 350°F.

Make incisions all over the meat. Into each one insert a sliver of garlic and a sprig of rosemary that you have dipped into a little salt. Stud the whole roast, then rub it with 2 tablespoons olive oil.

Scatter the carrots, whole clove garlics and few rosemary sprigs in the base of a roasting tin. Place the veal on a roasting rack, in the base of the pan. Roast in the oven for 1 hour 15 minutes.

Meanwhile, in the remaining oil, sauté the onion gently for about 20 minutes. Stir in the chopped garlic, then add the tomatoes and

raise the heat, cooking until the tomatoes melt into the onions. Add the mushrooms, reduce the heat and cook over medium-low heat, stirring occasionally, until the mushrooms are cooked through.

Remove the cooked veal from its pan, pour off any fat from the surface but save any juices. Add a few tablespoons of wine or stock. Place on the heat and scrape the base of the tin. Add the braised mushrooms and warm through, then set aside and keep warm.

Slice the roast, and serve each portion with a few spoonfuls of sauce.

Lamb with garlic and rosemary

pork noisettes with prunes and chestnuts

Peeling fresh chestnuts can be hard work but this method works every time.
If you prefer, use canned, ready-peeled, cooked chestnuts, available from most supermarkets.

SERVES 4

8 small pork noisettes

½ cup ready-to-eat prunes, chopped

⅔ cup red wine

⅔ cup chicken or vegetable broth

1 tablespoon liquid honey

2 tablespoons balsamic vinegar

½ cup chestnuts

2 tablespoons olive oil

Fresh rosemary sprigs and freshly
ground black pepper, to garnish

Wipe the noisettes and if necessary secure with fine twine or toothpicks to ensure they keep a good shape. Place in a shallow dish and scatter over the prunes. Mix together the red wine with the broth, honey, and vinegar and pour over the noisettes and prunes. Cover lightly and leave in the refrigerator for 30 minutes.

Make a slit at the top of the fresh chestnuts then boil for 10 minutes. Drain and dry thoroughly. (Omit this stage if you are using canned chestnuts.)

Lightly brush or spray a grill pan with oil then place on a moderate heat until hot. Add the chestnuts and cook for 10 to 15 minutes or until the skins begin to open. Remove

from heat, allow to cool, then peel. Set aside.

When ready to cook, lightly brush or spray the grill pan and heat as above then add 1 tablespoon of the oil. Drain the noisettes, reserving the marinade, and cook in the pan over a moderate heat for 4 to 6 minutes on each side or until done, adding the extra oil if the pan becomes dry.

Meanwhile place the marinade in a small pan and boil vigorously until reduced by half then stir in the chestnuts. Garnish the noisettes and serve with the chestnuts and sauce. Mustard-flavored mashed potatoes and fresh baby vegetables make the perfect accompaniments.

pan-fried pork with charred apple relish

While fresh cranberries are sour, dried ones are sweet but still retain a certain tartness which works well with pork. You can substitute raisins if dried cranberries are not available.

SERVES 4

Four 4-oz pork escalopes

2 tsp wholegrain mustard

½ cup clear apple juice

A few fresh sage leaves, lightly crushed

For the relish

8 oz apples, peeled, cored, and chopped fine

2 tablespoons dried cranberries or raisins

1 tablespoon cider vinegar

1 teaspoon wholegrain mustard

⅔ cup sour cream

Trim the escalopes if necessary, wipe, and place in a shallow dish. Blend the mustard with the apple juice and pour over the escalopes. Scatter with the sage leaves, cover lightly, and leave in the refrigerator for 30 minutes.

Lightly brush or spray a grill pan with oil then place on a moderate heat until hot. Add the apple pieces and cook for 2 to 3 minutes or until soft but not pulpy and slightly charred. Remove from the pan and add to the remaining relish ingredients, stir, and set aside.

Drain the escalopes and cook in the hot pan for 3 to 5 minutes each side or until the pork is done.

Slice the pork and place on a bed of spinach leaves. Serve with the apple relish, potatoes, and freshly cooked vegetables. Garnish with sage leaves.

calves' livers steamed in a herbed sauce

Fresh calves' livers are ideal for steaming. They remain moist and tender and, served in a mustard wine sauce, are a real treat. Try this recipe, especially if you think you've never liked liver. Full of iron, it's a highly nutritious meat.

SERVES 4

1 lb calves' livers

½ cup red wine

1 clove garlic, peeled and crushed

2 teaspoons fresh sage, chopped

1 leek, finely sliced

1 carrot, cut into julienne strips

For the sauce

1 tablespoon butter

1 tablespoon all-purpose flour

5 tablespoons plain yogurt or light cream

2 tablespoons fresh sage, chopped

Salt and freshly ground black pepper

Cut the livers into thin strips and place in a shallow glass dish. Mix the wine, garlic, and sage, and pour over the liver. Cover and marinate for 1 hour, turning occasionally.

Remove the liver, reserving the marinade, and place in a greaseproof paper-lined steamer along with the vegetables. Cover with a tight-fitting lid and steam for 10 minutes or until cooked through.

Meanwhile, melt the butter for the sauce in a small pan and stir in the flour. Remove from the heat and add the marinade, yogurt or cream and sage. Return to the heat and heat gently without boiling, stirring until thickened. Season to taste.

Place the liver and vegetables on a warmed plate, and top with the sauce. Garnish with sage and serve with boiled potatoes.

Pan-fried pork

sirloin steak with onion rings

Fried sage leaves combined with onion rings make a delicious topping for an exceptionally juicy steak. A simple mixture when combined with fries and a glass of cold beer.

SERVES 2

2 sirloin steaks, weighing 6 to 8 oz each

Salt and pepper

Oil, to brush

Topping

1 onion

8 sage leaves

2 teaspoons sea salt

4 teaspoons cornstarch

Oil, for deep frying

First prepare the topping. Peel and slice the onion into thin rings and discard the center. Sprinkle lightly with the salt and leave for 10 minutes, then rinse to remove the salt and pat dry with kitchen paper.

Dust the rings liberally with cornstarch and leave for 5 minutes, then turn them over, packing the cornstarch well down onto them.

Heat the oil in a deep pan until just smoking. Plunge the sage leaves into the oil until they crisp, but do not allow them to become very dark or they will taste bitter.

Lift them out and drain on kitchen paper. Plunge the onion rings into the oil for a couple of minutes, lift them out and while the oil is reheating, sprinkle them with a teaspoonful more of cornstarch.

Return them to the hot fat and fry them until they are really crisp. Drain on kitchen paper and keep warm.

Preheat the broiler, grill or barbecue. Season and oil the steaks and cook to your liking.
Serve with the crisp onion rings and sage leaves.

black olive tapenade

*This flavorful spread is delicious heaped onto crisp, thin little toasts, or spread
onto crusty baguettes for a heartier bite. Top with cream cheese or some fresh
mozzarella cheese. The tapenade is also good used as a relish with anything barbecued.*

SERVES 4

1 onion, chopped

4 tablespoons olive oil

3 ripe fresh tomatoes, shredded

1 lb fresh mushrooms, finely chopped

2 sprigs fresh thyme

Stock or white wine

3 garlic cloves, chopped

25 black olives, pitted and chopped

Ground black pepper, to taste

Lightly sauté the onion in olive oil until softened and golden brown, about 10 to 15 minutes.

Add the tomatoes, and continue cooking until the mixture becomes pastelike. Add the mushrooms and thyme, and cook gently for about 20 minutes, stirring and turning occasionally, until the mushrooms are very soft and tender. If the mixture becomes too dry, add some stock or white wine.

When the mixture has thickened, remove from the heat, and stir in the garlic and the olives. Mix well and season with pepper. Let cool to room temperature.

thyme

creamy leek and pasta flan

Whether served hot straight from the oven or chilled, this pasta flan tastes wonderful. The perfect contribution to a bring-a-plate party.

SERVES 6 TO 8

1½ cups orecchiette pasta

Dash of olive oil, plus 3 tablespoons

A little all-purpose flour

¾ lb packaged puff pastry, thawed if frozen

2 cloves garlic, minced

1 lb leeks, washed, trimmed, and cut into 1-in pieces

2 tablespoons chopped fresh thyme

2 eggs, beaten

⅔ cup light cream

Salt and freshly ground black pepper

1¼ cups shredded Cheddar cheese

Bring a large pan of water to a boil, and add the orecchiette with a dash of olive oil. Cook for about 10 minutes, stirring occasionally, until tender. Drain and set aside.

Dredge the work surface with flour and roll out the puff pastry to line a greased, 10-inch, loose-bottomed, fluted flan dish. For best results, chill in the refrigerator for at least 10 minutes.

Preheat the oven to 375°F. Heat the remaining olive oil in a large skillet and sauté the garlic, leeks, and thyme for about 5 minutes, stirring occasionally, until they become softened and tender. Then stir in the orecchiette, and continue to cook the mixture for a further 2 to 3 minutes.

Place the beaten eggs in a small bowl, then whisk in the cream, salt, and pepper. Transfer the leek and pasta mixture to the pastry case, spreading it out evenly. Pour the egg mixture over the top, then sprinkle with cheese. Bake for about 30 minutes, until the mixture is firm and the pastry crisp.

walnut and thyme risotto

This risotto is enriched with walnuts and their oil and the thyme acts as a delicate complement. Serve with a light green salad.

SERVES 4

5 cups vegetable stock

2 tablespoons butter

1 tablespoon olive oil

4 cloves garlic, minced

½ cup very finely chopped walnuts

2 tablespoons chopped fresh thyme or 2 teaspoons dried

2 cups arborio rice

Salt and freshly ground black pepper

1 tablespoon walnut oil

½ cup walnut pieces

Sprig fresh thyme, to garnish

Pour the stock into a pan and bring to a boil. Reduce the heat to a gentle simmer.

Meanwhile, melt the butter with the oil in a large pan and gently fry the garlic, chopped walnuts, and thyme for 2 minutes. Stir in the rice and cook, stirring, for a further 2 minutes until the rice is well coated in the walnut mixture.

Add the stock, ladle by ladle, until all the liquid is absorbed and the rice is thick, creamy, and tender. Keep the heat moderate. This will take about 25 minutes and should not be hurried.

Adjust the seasoning and stir in the walnut oil. Serve the risotto sprinkled with walnut pieces and garnish with thyme.

Creamy leek and pasta flan

spicy garlic shrimp

The silky sweetness of coconut milk combined with fiery spices makes this shrimp dish a perfect dinner party recipe. Reduce the quantity of chile used for a milder dish or simply remove the seeds, where most of the "heat" is found.

SERVES 4

2 cups uncooked shrimp, peeled and deveined

2 large zucchini, cut into julienne strips

1 red chile, finely chopped

1 carrot, cut into julienne strips

1 red bell pepper, cut into julienne strips

2 tomatoes, deseeded and chopped

2 tablespoons olive oil

1 teaspoon fresh gingerroot, grated

4 cloves garlic, peeled and crushed

Juice and zest of 1 medium lime

1 teaspoon ground turmeric

1 teaspoon coriander

1 teaspoon ground cumin

4 tablespoons coconut milk

1 tablespoon light soy sauce

8 oz dry egg noodles

Several sprigs of cilantro

Rinse the shrimp under running water and pat dry. Put in a shallow glass dish with the vegetables. Mix the oil, ginger, garlic, lime, spices, coconut milk and soy sauce together, and pour over the ingredients. Stir to coat, cover and marinate for 1 hour, turning occasionally.

Half-fill the base of a steamer with water and bring to a boil. Cover the base of the steamer top with dampened waxed paper and add the shrimp, vegetables, and marinade. Place over the steamer base, cover with a tight-fitting lid and steam for 10 minutes.

Put the noodles into the boiling water in the steamer base. Replace the shrimp and cook for a further 5 minutes or until the noodles and shrimp are cooked.

Remove the steamer top and drain the noodles. Arrange on a warmed serving plate and top with the shrimp mixture. Garnish with chopped cilantro, and serve.

Portuguese fish stew

Calling this a stew is a misnomer, as the cooking time is quite short. Make your own fish stock from fish trimmings, or buy fresh from the supermarket.

SERVES 4

1 onion, chopped

5 cloves garlic, minced

3 stalks celery, chopped

2 leeks, chopped

1 small bulb fennel, chopped

6 tablespoons olive oil

1 cup chopped tomatoes

2 teaspoons tomato paste

½ red bell pepper, cored, deseeded, and chopped

1 bay leaf

2-in strip orange zest

7½ cups fish stock

2¼ to 3 lb mixed shellfish and fish (except oily varieties), filleted

Large pinch of cayenne pepper

Salt and freshly ground black pepper

Cook the onion, garlic, celery, leeks, and fennel in oil for 45 minutes. Add the tomatoes, tomato paste, bell pepper, bay leaf, and orange zest. Cook briskly.

Add the fish stock, boil, then lower the heat. Add the fish and simmer for 40 minutes. Add cayenne pepper and seasoning to serve.

hummus

This Middle Eastern dip is delicious with fresh bread or crudites.
It is a classic use of garlic.

MAKES ABOUT 1 CUP

7-oz can garbanzo beans

3 plump cloves garlic, peeled

½ cup tahini paste

⅓ cup olive oil

Salt and freshly ground black pepper

Juice of ½ a lemon

Paprika

Drain the beans, reserving the liquid. Place the beans in a blender with the garlic, tahini and olive oil and blend until smooth.

Add as much liquid reserved from the beans as necessary to make a thick paste. Season well with the salt and pepper, then add lemon juice to taste.

Spoon the hummus into a serving dish and chill lightly. Sprinkle with paprika just before serving.

garlic chicken with cucumber and grapes

This dish has quite a delicate flavor. Serve with steamed rice and vegetables for a refreshing summer meal. Mango chunks would be ideal in place of grapes.

SERVES 4

4 skinned chicken breasts
   (about 4 oz each)

1 clove garlic, cut in half

Salt and freshly ground black pepper

1 cucumber, cut into julienne strips

4 oz seedless red grapes, halved

For the sauce

½ cup chicken stock

6 tablespoons light cream

2 tablespoons fresh tarragon, chopped

1 egg yolk

Rinse the chicken breasts under cold water and pat dry. Rub the garlic over both sides, season with salt and pepper and place in the bottom tier of a steamer with the garlic. Steam for 15 minutes.

Place the cucumber and grapes in the second steamer tier over the chicken and cook both tiers for 5 minutes or until chicken is cooked.

Meanwhile, mix all the sauce ingredients, except the egg, in a small pan and heat gently, but do not boil. Remove from the heat and whisk in the egg.

Remove the cucumber and grapes from the steamer and stir into the sauce. Transfer the chicken to warmed serving plates and serve with the sauce and fresh steamed vegetables.

tomato and basil dressing

A light, clean-tasting dressing for fish and shellfish, pasta, egg, chicken, or avocado salads. The basil enhances the flavor of the tomatoes. If necessary, add a little sugar to balance the flavor of the walnut oil.

MAKES ABOUT 1¼ CUPS

1 tablespoon olive oil

2 tablespoons walnut oil

2 tablespoons white wine vinegar

1 tablespoon sherry vinegar

3 well-flavored tomatoes

18 to 20 basil leaves, chopped

Dash of superfine sugar (optional)

Salt and freshly ground black pepper

Pour the oils and vinegars into a bowl. Whisk together.

Peel, deseed, and finely chop the tomatoes then stir into the dressing with the basil. Add a little sugar if necessary, then season to taste.

tarragon and sesame dressing

With its nutty taste, this dressing complements sliced, well-flavored tomatoes. In place of the sesame oil you can use walnut oil.

MAKES ½ CUP

4 tsp chopped fresh tarragon

1 tablespoon Dijon mustard

2 tablespoons lemon juice

2 tablespoons sesame oil

Dash of granulated sugar (optional)

Salt and freshly ground black pepper

Put all the ingredients into a bowl and whisk together. Make shortly before needed and leave in a cool place.

chive and lemon vinaigrette

Use this dressing to make a delicious potato salad by tossing it with warm potatoes, particularly new ones, and finely chopped scallions, and then allowing it to cool.

MAKES ABOUT ¾ CUP

1 clove garlic

Salt and freshly ground black pepper

Rind of 1 lemon, finely grated, and the juice

1½ teaspoon wholegrain mustard

4 tablespoons olive oil

2 tablespoons chopped chives

Put the garlic and a dash of salt into a bowl. Crush together, then stir in the lemon rind and juice, and the mustard until smooth.

Slowly pour in the oil, whisking constantly, until well emulsified.

Add the chives and season with black pepper.

salmoriglio

In Sicily, salmoriglio is used as the marinade for fish kabobs that are to be broiled or barbecued. Sicilians believe that the only way to make a really good salmoriglio is to add seawater. If you live miles from the shore, use sea salt instead!

MAKES ABOUT 1¼ CUPS

1 clove garlic

1 tablespoon finely chopped fresh parsley

1½ tsp chopped fresh oregano

About 1 teaspoon chopped fresh rosemary

Sea salt

¾ cup virgin olive oil, warmed slightly

3 tablespoons hot water

4 tablespoons lemon juice

Freshly ground black pepper

Put the garlic, herbs, and a dash of salt into a mortar or bowl and pound to a paste with a pestle or the end of a wooden spoon.

Pour the oil into a warm bowl then, using a fork, slowly pour in the hot water followed by the lemon juice, whisking constantly until well emulsified.

Add the garlic and herb mixture, and black pepper to taste. Put the bowl over a pan of hot water and warm for 5 minutes, whisking occasionally. Allow to cool before using.

HERBAL DRINKS

What better way to keep the chills at bay than to sip one of these winter warmer herbal drinks. Serve them piping hot.

lemon-balm wine cup

SERVES 10

1 bottle Bordeaux

1 small bunch of lemon balm

1 small bunch of borage

1 orange, sliced

½ cucumber, sliced thickly

1 liqueur glass of Cognac

1 tablespoon brown sugar

½ cup chilled soda water

Put all the ingredients except the soda water in a jug over ice for 1 hour, then stir well, strain and add the chilled soda water.

And it's just as good without the addition of the Cognac.

ginger and valerian tea

SERVES 4

1 teaspoon dried valerian

1 teaspoon grated fresh gingerroot

2 cups freshly boiled water

Juice of ½ lemon

2 to 3 teaspoons clear honey

Put the valerian and ginger in a tea pot or your chosen container for making herbal teas. Pour over the water and leave to brew for 3 minutes. Stir in the lemon juice and honey. Pour through a tea strainer.

chamomile and apple cup

SERVES 6

2 oz dried chamomile

2 cloves

3 green cardamoms

½ apple, unpeeled and sliced

1 cinnamon stick

Honey, light brown sugar, or maple syrup, to serve

Top far left lemon-balm wine cup; *Bottom far left* ginger and valerian tea; *Left* chamomile and apple cup

Put all the ingredients in a pan and bring to a boil. Lower the heat, stir and simmer for 5 minutes. Remove from the heat and pour into a cold container. Leave to stand for 5 minutes. Strain and pour into a jug or warmed tea pot to serve. The apple slices and cinnamon may be rinsed and added to the jug when the tea is served. Sweeten to taste with honey, light brown sugar, or maple syrup.

HERBAL DRINKS

Iced herbal teas make wonderfully refreshing drinks. Because chilling reduces the flavor, make double-strength tea, cool in the refrigerator and pour over ice to serve

apple, mint, and cranberry cooler

SERVES 4

4 tbsp dried cranberries

2 sprigs fresh mint

1 apple, peeled and sliced

¾ pt freshly boiled water

Put all the ingredients in a large tea pot, jug or other container and pour over the water. Leave to stand for 1 hour. Strain and chill before pouring over ice.

With alcohol 1½ tablespoons grenadine and 1 measure (1½ tablespoons) either gin or vodka per glass.

tomato and parsley sling

SERVES 4

2 cups tomato (or mixed vegetable) juice

1 small bunch parsley

Worcestershire or soy sauce to taste

Put the tomato or vegetable juice and parsley in a blender or food processor and blend. Pour over ice, and season with Worcestershire or soy sauce to taste.

With alcohol Add 1 measure (1½ tablespoons) vodka per glass.

nettle-ginger beer

MAKES APPROXIMATELY 10 GLASSES

4 oz fresh gingerroot, roughly chopped

3½ cups water

4 tablespoons superfine sugar

2 tablespoons dried nettles

2 teaspoons grated orange rind

1 cinnamon stick

Put the roughly chopped gingerroot and ½ cup water in a blender or food processor and blend. Dissolve the sugar in ½ cup of the water in a pan. Put all the ingredients in a jug or other container with a lid. Stir well and leave to stand in a cool place for

24 hours. Strain, put in the refrigerator to chill for about 1 hour and pour over ice to serve.

With alcohol Add 1 measure (1½ tablespoons) whiskey per glass.

Apple, mint, and cranberry cooler

resources and suppliers

To discover more about herbs and herbalism, the following contacts are recommended:

International Herb Association
910 Charles Street
Fredericksburg,
VA 22401, USA
www.iherb.org
In 1995 the IHA launched a promotional program featuring one group of herb plants each year. The week before Mother's Day was chosen as National Herb Week, preceded by a year-long educational effort for the Herb of the Year. The Herb of the Year in 2000 was Rosemary (*Rosmarinus*), 2001 featured Sage (*Salvia*), 2002 Coneflower (*Echinacea*), 2003 Basil (*Ocimum*), 2004 Garlic (*Allium sativum*), and 2005 Oregano/Marjoram (*Origanum*).

European Herbal Practitioners Association
Midsummer Cottage,
Nether Westcote, Oxon OX7 6SD, UK
www.users.globalnet.co.uk

Canadian Herb Society
5251 Oak Street,
Vancouver, BC, Canada V6M 4HI
www.herbsociety.ca

Australian Herb Society
PO Box 110, Mapleton,
Queensland 4560, Australia
www.maleny.net.au

HERBAL SUPPLIERS

Rio Grande Herb Co., Inc.
PO Box 12125
Albuquerque, NM 87195, USA
www.pagenism.com/ag/herbs

Creation Herbal Products
13765-C Hwy 221
Fleetwood, NC 28626, USA
www.creationherbal.com

G. Baldwin & Co Medical Herbalists
171/173 Walworth Road
London SE17 1RW, UK
www.baldwins.co.uk

East West Herbs
3 Neal's Yard
London WC2H 9DP, UK
www.eastwestherbshop.com

Global Herbal Supplies
Cnr Byrne & Eccles St., Cairns,
Mareeba, 4880 Queensland, Australia
www.globalherbalsupplies.com

picture credits

All images are copyright Quintet Publishing, except as follows:

pages 12, 13 Garden Exposures Photo Library;
page 14 Lynne Brotchie/Garden Picture Library;
page 16 John Glover/ Garden Picture Library;
page 18 Howard Rice/ Garden Picture Library;
page 20 Jerry Harpur/ Garden Picture Library/Garden Design Tessa Hobbs;
page 21 Jerry Harpur/ Garden Picture Library/Garden Design Chris Rosmini;
page 23 and 26 Marcus Harper/ Garden Picture Library/Garden Design HMP Leyhill;
page 27 Ron Sutherland/ Garden Picture Library;
page 29 Jerry Harpur/ Garden Picture Library/Garden Design Julia Scott;
page 30 Marcus Harpur/ Garden Picture Library;
page 31 top Andrew Lawson; bottom Garden Exposures Photo Library;
page 32 Christopher Fairweather/ Garden Picture Library ;
page 33 John Glover/ Garden Picture Library;
page 34 Diana Steedman;
page 35 John Glover/ Garden Picture Library;
page 41 Friedrich Strauss/Garden Picture Library;
page 42 Jerry Harpur/Garden Picture Library/Garden Design Joy Larkcom;
page 50 top A-Z Botanicals;
page 51 top Joy Michaud/Sea Spring Photos
page 52 top Sea Spring Photos
page 53 top A-Z Botanicals;
pages 55, 56, 59 Harry Smith Collection
page 60 top A-Z Botanicals;
page 61 Harry Smith Collection
page 62 top Michael Michaud/Sea Spring Photos;
page 65 top Sea Spring Photos;
page 68 top A-Z Botanicals;
page 69 top Garden Matters;
page 71 top Garden Exposures Picture Library;

page 72 top Garden Matters;
page 73 top A-Z Botanicals;
page 74 Peter McHoy
page 76 top A-Z Botanicals;
page 77 top Georgia Glynn-Smith/Garden Picture Library;
page 80 top Joy Michaud/Sea Spring Photos;
page 81 top Marcus Harpur/ Garden Picture Library;
page 83 A-Z Botanicals
page 84 top A-Z Botanicals;
page 88 top A-Z Botanicals;
page 91 top Garden Matters;
page 93 top, page 94 top Joy Michaud/Sea Spring Photos;
page 97 top Jerry Pavia/Garden Picture Library;
page 98 top A-Z Botanicals;
page 99 Peter McHoy
page 100 top Joy Michaud/Sea Spring Photos;
page 101 Harry Smith Collection
page 112 right, 113, 114 right, 122 top right, 130 top right, 134 top right, 136 bottom right, 138 top right Robert Harding Picture Library
page 140 top left Joy Michaud/Sea Spring Photos;
page 146 A-Z Botanicals;
page 149 Robert Harding Picture Library;
page 150 Garden Matters;
page 151 top, Robert Harding Picture Library; bottom ET Archive;
page 152, 154, 160, 162 left, 163 bottom, 165 Robert Harding Picture Library;

Illustrations by Nicola Gregory, Elisabeth Dowle, Sally Launder, and Sharon Smith
Additional photography by Ian Garlick, Jeremy Thomas, Paola Zucchi, Tim Ferguson Hill, Nelson Hargreaves, Chas Wilder, Paul Forrester, Keith Waterton, Ian Howes.
Recipes were contributed by Gina Steer, Rosemary Moon, Jenny Stacey, Marlena Speiler, and Kathryn Hawkins

index

a

Achillea filipendulina **49**

Achillea millefolium **21**, **49**

    A. millefolium var. *rosea* **49**

Aconitum napellus **30**

After-bath Cologne, Floral **114**

After-bath Oil, Herbal **127**

After-sun Moisturizer **113**

Ajuga reptans **26**

    A. reptans Atropurpurea **30**

    A. reptans Burgundy Glow **30**

Alcea rosea **24**, **30**

Alchemilla mollis **22**, **29**, **50**

A. vulgaris **50**

Allium sativum **42**, **51**, **144**

Allium schoenoprasum **15**, **52**

Alo barbadensis **53**, **146**

aloe vera **53**

    in cosmetics **113**

    medicinal use **146**

Aloysia triphylla **15**, **54**

Althaea officinalis **55**

Anethum graveolens **15**, **56**

angelica **24**, **25**, **29**, **40**, **57**

    Chinese **57**

    medicinal use **147**

    to sow **37**

Angelica archangelica **25**, **29**, **57**

    A. acutiloba **147**

    A. atropurpurea **57**

    A. polymorpha var. *sinensis* **57**

    A. sinensis **147**

Anthemis nobilis **58**

Anthriscus cerefolium **59**

Apple, Mint, and Cranberry Cooler **217**

arnica **60**

Arnica montana **60**

Artemisia abrotanum **28**, **42**, **61**

    A. absinthum **28**

A. absinthum Lambrook Silver **28**

A. dracunculoides **15**, **62**

A. dracunculus **15**, **62**

A. Powis Castle **28**

A. stelleriana Boughton Silver **28**

A. vulgaris **42**

Astringent, Marigold **117**

b

Baked Lasagne with Chervil **175**

basil, sweet **15**, **85**, **167**

    in cosmetics **126–7**

    to freeze **44**

    purple **15**, **30**

    recipes **188–90**, **212**

Bath Salts, Herbal **119**

Bath Oil, Rosemary **135**

bay **14**, **77**, **167**

    cuttings **40**

    to dry **44**

    yellow-leaved **29**

bergamot **30**, **81**

    to divide **39**

    lemon **81**

    red **81**

    wild **81**

bergamot orange **21**

Black Olive Tapenade **205**

Body Lotion, Lavender **121**

Body Oil, Energizing **125**

Body Toner, Spicy **141**

borage **63**

    to sow **37**

Borago officinalis **63**

Bruise Treatment, Parsley **131**

bugle **26**, **30**

c

Calves' Livers Steamed in a Herbed Sauce **202**

caraway **65**

cardinal flower **30**

Carum carvi **65**

catmint **26**, **83**

    in cosmetics **112**

    to divide **39**

    to prune **43**

Centranthus ruber **31**, **102**

Chamaemelum nobile **21**, **22**, **32**

    C. nobile Flore pleno **22**

    C. Treneague **32**

chamomile **32**, **41**, **58**

    in cosmetics **115**, **137**, **139**

    lawn **32**

    Roman **21**, **22**, **32**, **58**

Chamomile and Apple Cup **215**

Charred Cod with Pesto **188**

chervil **42**, **59**

    in cosmetics **139**

    recipes **175–6**

    to sow **37**

Chicken with Cucumber and Grapes, Garlic **211**

Chicken Dumplings with Chives **170**

Chicken and Tarragon Mushroom Risotto **178**

Chicken with Tarragon, Pan-fried **177**

Chive and Lemon Vinaigrette **213**

chives **15**, **52**, **59**, **167**

    to divide **39**

    to freeze **44**

    recipes **168–70**, **213**

Chrysanthemum parthenium **23**, **47**, **66**

    see also *Tanacetum parthenium*

cilantro **15**, **67**
    recipes **180–2**
    to sow **37**
Citrus bergamia **21**
clary **95**
Cleanser, Parsley and Mint **131**
    Yarrow and Nettle **112**
Cleansing Face Pack **112**
Cleansing Cream, Comfrey **139**
    Lavender **121**
Cod with Pesto, Charred **188**
comfrey **24**, **25**, **99**
    in cosmetics **112**, **138–9**
    to divide **39**
    to prune **43**
companion planting **42–3**
Conditioner, Superb Elder **137**
Conditioning Night Cream **129**
cone flower **24**, **30**, **41**, **69**
    see also Echinacea
Consolida ambigua **25**
coriander **67**
Coriandrum sativum **15**, **67**
cosmetics: herbs for **18–21**
    recipes **104–141**
cotton lavender **28**, **42**
    to prune **43**
cranesbill **26**, **43**
Crataegus laevigata **68**
    C. oxyacantha **148**
Cream, Cleansing **121**, **139**
    Conditioning Night **129**
    Healing Hand **129**
    Marigold **117**
    Mint Skin **123**
    Rose and Apricot **133**
Creamy Leek and Pasta Flan **207**
curry plant **43**
Cynara cardunculus Scolymus group
    28

d

dandelion **112**
Deodorant, Lavender **121**
Dianthus **28**
Digitalis purpurea **30**
dill **15**, **56**, **167**
    recipes **171–3**
    to sow **37**
dressings **212–13**
drinks **215–17**
drying herbs **44**

e

Echinacea **69**
    E. angustifolia **25**
    E. purpurea **25**, **30**, **150**
E. purpurea Leuchstern **30**
E. purpurea White Swan **25**
    medicinal use **150–1**
Eggplant, Fennel, and Walnut Salad **187**
Eggplant Pesto **190**
elder **21**, **96**, **136**
    American **21**
    golden **29**
elderflower: in cosmetics **112**, **131**,
    136–7, **139**
    vinegar **107**
water **21**, **106**
Eleutherococcus senticosus **152**
    see also Panax ginseng
Energizing Body Oil **125**
Eryngium giganteum Silver Ghost **28**
    E. maritinum **28**
    E. maritinum Miss Wilmott's
    Ghost **28**
Erysimum asperum Bowles Mauve **31**
    E. cheiri **31**
Eucalyptus **28**
evening primrose **24**, **25**, **84**

Eye Lotion, Fennel **118**
    Mint **123**

f

Face Freshener, Elderflower **137**
Face Mask, Basil and Lemon **127**
    Comfrey **139**
    Peppermint **125**
    Sweet Leaf **129**
Facial, Fennel and Olive Oil **118**
    Rosemary and Almond **135**
Fattoush **195**
fennel **15**, **21**, **70**
    bronze **21**, **30**
    in cosmetics **112**, **118**
    recipes **184–7**
    to sow **37**
fenugreek **101**
feverfew **23**, **47**, **66**
    in cosmetics **112**
    golden **28**
    medicinal use **162**
    to sow **37**
Fish Stew, Portuguese **208**
Floral After-bath Cologne **114**
Floral Skin Tonic **135**
Foeniculum vulgare **15**, **21**, **70**
    F. vulgare Purpureum **21**, **30**
Foot Balm, Lavender **121**
foxglove **30**
Freckle Lotion, Parsley **131**
Fried Parsley and Celery Leaves **196**

g

garlic **42**, **51**
    in cosmetics **139**
    medicinal use **144–5**
    oil **44**

recipes **208–11**
Garlic Chicken with Cucumber and
Grapes **211**
Geranium **26**
G. Johnson's Blue **26**
G. magnificum **26**
geranium, rose **114**, **128–9**, **139**
sweet-leaved **89**
ginger **103**
medicinal use **164**
Ginger and Valerian Tea **215**
gingko **71**
medicinal use **154–5**
Gingko biloba **71**, **154**
ginseng **88**
medicinal use **152–3**
globe artichoke **28**
Glycyrrhiza glabra **72**, **156**
Goat Cheese Tartina **168**
golden rod **25**
Green Healing Mask **131**
gum tree **28**

h

Hair Rinse, Elderflower **137**
Hamamelis virginiana **24**
Hand Cream, Healing **129**
harvesting herbs **44**
hawthorn **68**
medicinal use **148–9**
Healing Hand Cream **129**
Healing Mask, Green **131**
herb garden: to maintain **34–40**
to plant **34–40**
Herbal After-bath Oil **127**
Herbal Bath Salts **119**
hollyhock **24**, **30**
honeysuckle **11**, **43**
hop **43**

golden **29**
Hot and Sour Soup **180**
houseleek **23**, **31**
hummus **210**
Humulus lupulus Aureus **29**
Hypericum perforatum **73**, **157**
hyssop **74**
cuttings **40**
Hyssopus officinalis **74**

i

Iris germanica **75**
Iris Florentina **21**

j

jasmine **21**, **76**
Arabian **76**
in cosmetics **114**
Spanish **21**, **76**
Jasminum grandiflorum **21**, **76**
J. officinalis **21**, **76**
J. sambac **76**
Jelly, Marigold **117**

k

kava kava **91**
medicinal use **159**

l

lad's love *see* southernwood
lady's mantle **11**, **22**, **28**, **50**
in cosmetics **112**, **139**
to prune **43**
to sow **37**
Lamb Steaks with Garlic and Rosemary
199

larkspur **25**
Laurus nobilis **14**, **77**
L. nobilis Aurea **29**
Lavandula angustifolia **18**, **20**, **78**
L. stoechus f. *leucantha* **28**
lavender **11**, **18**, **20**, **21**, **28**, **42**, **78**
in cosmetics **119**, **120–1**, **122**,
135, **137**
cuttings **40**
to dry **44**
Hidcote Pink **20**
Imperial Gem **20**
Munstead **20**
Nana Alba **20**
to prune **43**
water **106**
Leek and Pasta Flan, Creamy **207**
lemon balm **22**, **79**
in cosmetics **135**
golden **28**
to prune **43**
Lemon Balm Wine Cup **215**
lemon verbena **15**, **54**
in cosmetics **114**
licorice **72**
medicinal use **156**
Lightening Paste, Chamomile **115**
Lilium candidum **20**
Linaria alpina **31**
Lip Balm, Rose **133**
Lobelia cardinalis **30**
Lotion, Basil and Rosewater **127**
Elderflower **137**
Lavender Body **121**
Massage **113**
Parsley Freckle **131**
Rosemary Herbal **135**
Thyme and Lemon **141**
lungwort **26**

m

Madonna lily **20**, **129**

maidenhair tree *see* gingko

marigold **11**

    in cosmetics **116–17**, **139**

    French **25**, **42**

    pot **17**, **47**, **64**

    to sow **37**

marjoram, sweet **15**, **87**

    to divide **39**

    golden **28**

    pot **15**

    to prune **43**

marshmallow **55**

Mask, Basil and Lemon Face **127**

    Comfrey, Face **139**

    Green Healing **131**

    Peppermint Face **125**

    Rose Minute **133**

    Sweet Leaf Face **129**

    Thyme and Fig **141**

Massage Lotion **113**

medicinal herbs: to grow **22–25**

    to use **143–65**

Melissa officinalis **22**, **79**

    M. officinalis Allgold **28**

Mentha **80**

    M. piperita **22**, **80**, **158**

    M. x piperita citrata **30**

M. spicata **15**

milk thistle **24**, **25**, **98**

    medicinal use **161**

mint **12**, **15**, **40**, **80**

    basil **15**

    in cosmetics **122–3**, **141**

    eau de Cologne **15**, **30**

    to freeze **44**

    ginger **15**

    pineapple **15**

    to prune **43**

    see also peppermint, spearmint

Moisturizer, Rose **133**

    After-sun **113**

Moisturizing Milk, Mint and Parsley **123**

Monarda citriodora **81**

    M. didyma **30**, **81**

    M. fistulosa **81**

monkshood **30**

Mouthwash, Mint and Rosemary **123**

    Peppermint **125**

mugwort **43**

mullein **24**, **28**

Mussels with Creamy Fennel Sauce, Steamed **184**

Myrrhis odorata **82**

n

nasturtium **17**, **41**

Nepeta cataria **26**, **83**

Nettle-Ginger Beer **217**

Night Cream, Conditioning **129**

o

Oenothera biennis **25**, **84**

Ocimum basilicum **85**

    O. basilicum Dark Opal **30**

    O. basilicum Purple Ruffles **15**, **30**

oregano **15**, **86**

    recipes **191–2**, **213**

Oregano onites **15**

Oreganum majorana **15**, **87**

    O. vulgare **15**, **86**

    O. vulgare Aureum **28**

orris root **21**, **75**

    in cosmetics **119**

p

Panax ginseng **88**

    see also Eleutherococcus senticosus

Pan-fried Chicken with tarragon **177**

Pan-fried Pork with Charred Apple Relish **202**

Pan-fried Salmon with Black Bean Relish **182**

parsley **15**, **59**, **90**, **167**

    in cosmetics **123**, **130–1**

    to freeze **44**

    Neapolitan **90**

    recipes **194–6**, **213**, **217**

    to sow **37**

pasta: Baked Lasagne with Chervil **175**

    Creamy Leek and Pasta Flan **207**

    Pasta Primavera **196**

    Salmon and Broccoli Penne **172**

Paste, Marigold and Yogurt **117**

Pelargonium **89**

    P. crispum **17**, **89**

    P. fragrans **89**

    P. graveolens **17**, **21**, **89**

    P. odoratissimum **89**

    P. quercifolium **89**

    P. radens **89**

    P. tormentosum **17**, **89**

peppermint **22**, **80**

    in cosmetics **124–5**, **126**

    medicinal use **158**

Perfume, Rose and Basil **127**

Petroselinum crispum **15**, **90**

Phlomis fruticosa **29**

pinks **28**

    Mrs Sinkins **28**

Piper methysticum **91**, **159**

Pork with Charred Apple Relish, Pan-fried **202**

Pork Noisettes with Prunes and
     Chestnuts **200**
Portuguese Fish Stew **208**
Potato Salad, Russian Dilled **171**
Poudre à la Mousseline **119**
Poultice, Comfrey and Garlic **139**
preserving herbs **44**
propagation **37–40**
pruning **43**
Pulmonaria officinalis **26**
     P. officinalis Cambridge Blue **26**

r

Ratatouille and Goat Cheese Quiche
     189
Risotto alla Bolognese **191**
Risotto, Chicken and Tarragon
     Mushroom **178**
     Scallop and Fennel **187**
     Smoked Salmon and Dill **173**
     Walnut and Thyme **207**
Roasted Portobellos with Pine Nuts and
     Chervil **176**
Rosa **92**
     R. x alba **18**
     R. x alba Semiplena **18**
     R. x centifolia Muscosa **18**
     R. x damascena var.
     sempiflorens **18**
     R. gallica var. *officinalis* **18**
     R. gallica Versicolor **18**
rose **18, 21, 92**
     alba **18**
          Celeste **18**
          Königen von Dänemark **18**
          Maiden's Blush **18**
     White Rose of York **18**
Apothecary's **11, 18**
Bourbon **18**

Louise Odier **18**
Madame Isaac Pereire **18**
Zephirine Drouhin **18**
in cosmetics **114, 135, 137**
damask **18**
     Autumn Damask **18**
     Professeur Emile Perrot **18**
to dry **44**
gallica **18**
     Duc de Guiche **18**
     Rosa mundi **18**
     Tuscany Superb **18**
moss **18**
     Chapeau de Napoleon **18**
     Cristata **18**
     Hunslet Moss **18**
to prune **43**
vinegar **107**
rose water **106, 127, 132–3, 135, 137**
rosemary **11, 14, 21, 22, 28, 42, 93,**
     167
     in cosmetics **122, 123, 134–5, 141**
     cuttings **40**
     Miss Jessop's Upright **14**
     Pinkie **20**
     to prune **43**
     recipes **199–200, 213**
     Texas rosemary Arp **20**
Rosemary-roasted Veal **199**
Rosmarinus officinalis **14, 20, 22, 93**
     R. officinalis var. *albiflorus* **28**
     R. officinalis Aureus **20**
     R. officinalis Prostratus group **14**
rue **11**
Russian Dilled Potato Salad **171**

s

sage **14, 20, 22, 28, 94**

in cosmetics **112, 123**
cuttings **40**
golden **28**
Jerusalem **29**
pineapple **17**
to prune **43**
purple **30**
recipes **202–4**
Spanish **20**
St John's wort **73**
medicinal use **157**
Salmon with Black Bean Relish, Pan-
     fried **182**
Salmon and Broccoli Penne **172**
Salmoriglio **213**
Salvia elegans Scarlet Pineapple **17**
S. lavandulifolia **20**
S. officinalis **14, 20, 22, 94**
S. officinalis Albiflora **28**
S. officinalis Icterina **14, 28**
S. officinalis Kew Gold **22, 28**
S. officinalis Purpurescens group **14**
S. purpurea Rasberry Royal **30**
     S. sclarea **95**
Sambucus canadensis **21, 96**
     S. nigra **21, 96**
     S. nigra Aurea **21, 29**
     S. nigra Guincho Purple **21**
     S. nigra f. *laciniata* **21**
     S. racemosa Plumosa Aurea **29**
San Francisco Cioppino **192**
Santolina chamaecyparissus **28**
saw palmetto **97**
     medicinal use **160**
Scallop and Fennel Risotto **187**
Scallops in Saffron Cream **169**
sea holly **28**
Sempervivum **23, 31**
Serenoa repens **97, 160**
Shrimp, Spicy Garlic **208**

Silybum marianum 25, 98, **161**
Sirloin Steak with Onion Rings **204**
Skate with Tarragon Butter **178**
Skin Cream, Mint **123**
Skin Tonic, Chamomile **115**
    Floral **135**
    Rose Geranium **129**
Smoked Haddock Pots **194**
Smoked Salmon and Dill Risotto **173**
Soap Balls, Thyme **141**
Solidago virgaurea **25**
southernwood 28, **42**, 61
    cuttings **40**
    to prune **43**
spearmint **15**, 80
Spicy Body Toner **141**
Spicy Garlic Shrimp **208**
Steamed Mussels with Creamy Fennel
    Sauce **184**
Stinking Roger **42**
Sun Oil, Lemon Verbena **114**
Superb Elder Conditioner **137**
sweet cicely **82**
Sweet Leaf Face Mask **139**
Symphytum officinale 25, **99**

t

Tagetes patula **25**
    T. minuta **42**
Tanacetum argentum **28**
    T. parthenium 23, **162**
    T. parthenium Aureum **28**
    see also Chrysanthemum
    parthenium
tansy 28, **112**
Tapenade, Black Olive **205**
tarragon, French 12, 15, 59, 62, **167**

to prune **43**
recipes 177–8, **212**
Russian **15, 62**
Tarragon and Sesame Dressing **212**
Teuchrium chamaedrys **31**
thyme 14–15, 20, 32, **100**
    caraway **100**
    in cosmetics 140–1
    golden **28**
    lemon **100**
    to prune **43**
    recipes 205–7
    Silver Posie **15**
Thymus citriodorus **100**
    T. x citriodorus Argentius **28**
    T. drucei **100**
    T. herba-barona **100**
    T. serpyllum Annie Hall **32**
    T. serpyllum Pink Chintz **32**
    T. serpyllum Rainbow Falls **32**
T. vulgaris 14–15, 20, **100**
toadflax, alpine **31**
Tomato and Basil Bruschetta **190**
Tomato and Basil Dressing **212**
Tomato and Cilantro Soup **181**
Tomato and Parsley Sling **217**
Toner, Comfrey **139**
    Peppermint Lemon **125**
    Spicy Body **141**
Trigonella foenum-graecum **101**
Tropaeolum majus **17**

V

valerian **31, 102**
    medicinal use **163**
Valeriana officinalis 30, 102, **163**
Veal, Rosemary-roasted **199**

Verbascum thapsus 24, 28
Viola odorata 21
violet, sweet **21**

W

wallflower **31**
wall germander **31**
Walnut and Thyme Risotto **207**
witch hazel **24**
    in cosmetics 125, 133, **137**
wormwood 11, 28, **43**

y

yarrow 21, 42, **49**
    in cosmetics **112**
    to divide **39**

z

Zingiber officinale 103, **164**